全国高职高专规划教材

环境工程专业英语

English for Environmental Engineering

张之浩　王　晖　主编

中国环境出版集团·北京

图书在版编目（CIP）数据

环境工程专业英语 / 张之浩，王晖主编．—北京：中国环境出版集团，2014.8
（2022.8 重印）
全国高职高专规划教材
ISBN 978-7-5111-2021-2

Ⅰ.①环… Ⅱ.①张… ②王… Ⅲ.①环境工程—英语—高等职业教育—教材 Ⅳ.① H31

中国版本图书馆 CIP 数据核字（2014）第 162786 号

出 版 人　武德凯
责任编辑　黄晓燕　侯华华
责任校对　薄军霞
封面设计　宋　瑞

出版发行　中国环境出版集团
　　　　　（100062 北京市东城区广渠门内大街 16 号）
　　　　　网　　址：http://www.cesp.com.cn
　　　　　电子邮箱：bjgl@cesp.com.cn
　　　　　联系电话：010-67112765（编辑管理部）
　　　　　　　　　　010-67112735（第一分社）
　　　　　发行热线：010-67125803，010-67113405（传真）
印　　刷　北京中科印刷有限公司
经　　销　各地新华书店
版　　次　2014 年 8 月第 1 版
印　　次　2022 年 8 月第 3 次印刷
开　　本　170×230
印　　张　13.5
字　　数　340 千字
定　　价　33.00 元

【版权所有。未经许可，请勿翻印、转载，违者必究。】
如有缺页、破损、倒装等印装质量问题，请寄回本社更换

《环境工程专业英语》编委会

主　审：谌桂君
主　编：张之浩　王　晖
参　编：周建华　李　欢　王真真　杨　帆
　　　　易彩纯　蔡　媛　刘　峥

前　言

教育部高等教育司2006年颁布的《高职高专教育英语课程教学基本要求（试行）》中指出："在完成《基本要求》的教学任务后，应结合专业学习，开设专业英语课程，这既可保证学生在校期间英语学习的连续性，又可使他们所学的英语得到实际的应用。"这说明为培养面向21世纪的高等技术应用性人才，开设专业英语是十分必要的。

由于环境工程专业具有多学科性和全球性的特点，对于本专业的高等技术应用性人才，在以英语为工具，获取专业所需信息，处理与实际工作有关的英语科技资料以及进行基础的涉外专业交流等方面，应具有更强的能力。为了达到这一目标，长沙环境保护职业技术学院专职英语教师和环境工程专业教师合作编写了《环境工程专业英语》教材，供相关专业的高职高专学生试用，并在教学实践中修订与完善。

本书具有如下特点：

（1）专业内容选用最新科技材料，重点突出，要点兼顾。全书共分四个部分，含10个单元，每单元编有课文一篇或两篇，配练习题若干，部分单元附有阅读文章对课文内容进行补充。内容涉及环境科学、生态学、环境工程和环境污染等方面，重点突出水、气、渣的污染治理和资源回收利用及分析检测技术等环境工程专业知识。

（2）本书体裁多样，语言规范，实用性强。选材时考虑到了未来的环境工程技术人员在实际工作中常涉及的英语资料的主要方面，材料的实用性成为编者选材的重要标准，尽可能选取了相关专业最基础和最广泛应用的专业语汇和专业材料来编写词汇表、注释和习题，并将水处理平台的操作流程编为教学重点。体裁涵盖了技术论文、科技报告、检测方法和实验器材说明等。

（3）本书主要读者对象为大专院校和职业技术学院环境工程相关专业学生及相关专业技术人员。编写时除了考虑到内容的连续性、相关性和全书的梯度外，还对原文中较专业的内容及较难的词和句进行了注释。各单元后附精心编排的习题、习题答案和课文参考译文也是本书的一大特点，既便于教师组织课堂教学和操练，又为学生和自学者提供了充足的自修材料。练习题的题型多样，主要有简答问题、多项选择和正误判断——旨在提高学生的阅读理解能力；图表填空、词和短语的翻译和匹配——促使学生熟记本专业常用的词汇和表达法；现场模拟阐述——旨在锻炼学生的专业英语口语表达能力；英译汉——提高学生对长、难句的理解能力；汉译英——培养学生专业英语方面的初步写作能力；段落翻译——内容与本单元有关，但难度稍低，可用来测试学生的阅读理解能力。

本教材编写组由长沙环境保护职业技术学院环境工程系三位专业教师和基础部五位专职英语教师组成，张之浩、王晖任主编，第一部分第一单元、第五单元和第二部分第一单元、第二单元由张之浩、王晖编写，第一部分第二单元由李欢、蔡媛编写，第一部分第三单元由李欢、杨帆编写，第一部分第四单元由王真真、周建华编写，第二部分第一单元、第二单元由王真真、易彩纯编写，第三部分第一单元、第二单元由王真真、王晖编写，第四部分由刘峥编写。教材课文选材由张之浩、李欢和王真真负责完成，课文后的生词音标、表达法解释、课文注释、课后习题、习题答案和课文翻译由王晖、周建华、易彩纯、杨帆、蔡媛负责完成，其中张之浩提供的水处理平台课文系原创。教材后附的专业词汇表由李欢提供。本教材体现了高职英语教学为学生专业服务的宗旨，将由专职英语教师和环境工程专业教师在合作授课的教学实践中根据学生实际水平和行业要求不断修订和完善。

本教材在编写和出版过程中得到了长沙环境保护职业技术学院李倦生书记、吕文明书记、刘益贵院长、黄忠良副院长、孙蕾副院长、曾桂华副院长、环境工程系吴同华主任和教务处刘杨林处长等领导的大力支持，他们提出了许多宝贵意见，谨在此表示衷心感谢。

本教材涉及环境工程专业内容较多，学科面广，以及限于编者水平，错误之处在所难免，希望读者不吝指正，不胜感谢。

【本书系湖南省教育科学"十二五"规划2012年度大学英语教学研究专项课题"高职理工科学生翻译技能培养有关问题研究"（课题批准号：XJK12YYB003)与2012年度湖南省教育厅科学研究项目一般项目"对科技翻译研究困境的反思及对策研究"（项目批准号：12C0927）的部分研究成果。也是长沙环境保护职业技术学院2017年项目化教学改革课程《环保专业英语1》的前期研究成果。】

编者
2018年7月

Content

Part I Water Treatment 1

Unit 1 Brief Introduction to Water Treatment 1
        Text Water Pollutants and Their Sources 1

Unit 2 Primary Treatment 11
        Text A Wastewater Pretreatment (1) 11
        Text B Wastewater Pretreatment (2) 19

Unit 3 Secondary Treatment 27
        Text A Aerobic Treatment - Activated Sludge System 27
        Text B Anaerobic Wastewater Treatment 36
        Text C Flocculation 44

Unit 4 Tertiary Treatment 50
        Text A Applications of Advanced Oxidation 50
        Text B Advanced Waste Treatment of Secondary Effluent with
                Active Carbon 59

Unit 5 Sewage Treatment Plant 72
        Text A Typical Project of Sewage Treatment Plant (1) 72
        Text B Typical Project of Sewage Treatment Plant (2) 82

Part II Air Pollution Control Techniques 89

Unit 1 Air Pollution Control (Gas) 89
        Text Control of Gas 89

  Unit 2 Air Pollution Control (Particulates) ………………………………… 97
    Text Control of Particulates ……………………………………… 97

Part III Solid Waste Management ……………………………………… 111

  Unit 1 Introduction to Solid Waste Management………………………… 111
    Text Solid Waste ……………………………………………… 111
  Unit 2 Summary of Solid Waste Management…………………………… 119
    Text Integrated Waste Management …………………………… 119

Part IV Abstract Writing ……………………………………………… 129

附录 ………………………………………………………………………… 137

  附1：环境工程专业用语参考 …………………………………………… 137
  附2：参考译文与参考答案 ……………………………………………… 149

Part I Water Treatment

Unit 1 Brief Introduction to Water Treatment

Text Water Pollutants and Their Sources

The wide range of pollutants that are being discharged to surface waters can be grouped into broad classes.

Point Sources

Domestic sewage and industrial wastes are called point sources because they are generally collected by a network of discharge into the receiving water of pipes or channels and conveyed to a single point of discharge into the receiving water. Domestic sewage consists of wastes from homes, schools, office buildings, and stores. The term （术语） municipal sewage is used to mean domestic sewage into which industrial wastes are also discharged. In general, point source pollution can be reduced or eliminated through proper wastewater treatment prior to discharge to a natural water body.

Non-point Sources

Urban and agricultural runoff are characterized by multiple discharge points. These are called non-point sources. Often flow of polluted water flows over the surface of the land or along natural drainage channels to the nearest water body. Even when urban or agricultural runoff waters are collected in pipes or channels, they are generally transported the shortest possible distance for discharge, so that wastewater treatment at each outlet is not economically feasible. Much of the non-point source pollution occurs during rain storms or spring snowmelt resulting in large flow rates that make treatment even more difficult. Reduction of non-point source pollution generally requires changes in land use practices and improved education.

Oxygen-demanding Material

Anything that can be oxidized in the receiving water with the consumption of dissolved molecular oxygen is termed oxygen-demanding material. This material is usually biodegradable organic matter but also includes certain inorganic compounds. The consumption of dissolved oxygen, DO, poses a threat to higher forms of aquatic life that must have oxygen to live. The critical level of DO varies greatly among species. For example, brook trout （溪点红蛙） may require about 7.5 mg/L of DO, while carp （鲤鱼） may survive at 3 mg/L. As a rule, the most desirable commercial and game fish require high levels of dissolved oxygen. Oxygen-demanding materials in domestic sewage come primarily from human waste and food residue. Particularly noteworthy among the many industries which produce oxygen-demanding wastes are the food processors and the paper industry. Almost any naturally occurring organic matter, such as animal droppings, crop residues, or leaves, which get into the water from non-point sources, contribute to the depletion of DO.

Nutrients

Nitrogen and phosphorus, two nutrients of primary concern, are considered pollutants because they are too much of a good thing. All living things require these nutrients for growth. Thus, they must be present in rivers and lakes to support the natural food chain. Problems arise when nutrient levels become excessive and the food web is grossly disturbed, which causes some organisms to proliferate at the expense of others. As will be discussed in a later section, excessive nutrients often lead to large growths of algae, which in turn become oxygen-demanding material when they die and settle to the bottom. Some major sources of nutrients are phosphorus-based detergents, fertilizers, and food-processing wastes.

Pathogenic Organisms

Microorganisms found in wastewater include bacteria, viruses, and protozoa excreted by diseased persons or animals. When discharged into surface waters, they make the water unfit for drinking. If the concentration of pathogens is sufficiently high, the water may also be unsafe for swimming and fishing. Certain shellfish（甲壳类动物，贝类）can be toxic because they concentrate pathogenic organisms in their tissues, making the toxicity levels in the shellfish much greater than the levels in the surrounding water.

Suspended Solids

Organic or inorganic particles that are carried by the wastewater into a receiving water are termed suspended solids (SS). When the speed of the water is reduced by flowing into a pool or a lake, many of these particles settle to the bottom as sediment. In common usage, the word sediment also includes eroded soil particles which are being carried by water even if they have not yet settled. Colloidal particles which do not settle readily cause the turbidity found in many surface waters. Organic suspended solids may also exert an oxygen demand. Inorganic suspended solids are discharged by some industries but result mostly from soil erosion that is particularly bad in areas of logging, strip mining（露天开采）, and construction activity. As excessive sediment loads are deposited into lakes and reservoirs, their usefulness is reduced. Even in rapidly moving mountain streams, sediment from mining（采矿业）and logging operations（采伐作业）has destroyed many living places (ecological habitats) for aquatic organisms. For example, salmon（鲑鱼）eggs can only develop and hatch in loose gravel（松散的砾石）stream beds. As the pores（气孔）between the pebbles（鹅卵石）are filled with sediment, the eggs suffocate（窒息）and the salmon population is reduced.

Toxic Metals and Toxic Organic Compounds

Agricultural runoff often contains pesticides and herbicides that have been used on crops. Urban runoff is a major source of lead and zinc in many water bodies. The lead comes from the exhaust of automobiles using leaded gasoline, while the zinc comes from tire wear（轮胎磨损）. Many industrial wastewaters contain either toxic metals or toxic organic substances. If discharged in large quantities, many of these materials can make a body of water nearly useless for long periods of time. The lower James River in Virginia has been reduced to use only as a shipping channel because of a large industrial discharge of highly toxic and persistent organic compounds. Many toxic compounds are concentrated in the food chain, making fish and shellfish unsafe for human consumption. Thus, even small quantities in the water can be incompatible（不相容的）with the natural ecosystem and many human uses.

Heat

Although heat is not often recognized as a pollutant, those in the electric power industry are well aware of the problems of disposing of waste heat. Also, many industrial process waters are much hotter than the receiving waters. In some environments an increase of water temperature can be beneficial. For example, production of clams（蛤蚌）and oysters（牡蛎）can be increased in some areas by warming the water. On the other hand, increases in water temperature can have negative impacts. Many important commercial and game fish such as salmon and trout（鳟鱼）will only live in cool water. In some instances the discharge of heated water from a power plant can completely block salmon migration（洄游）. Higher temperatures also increase the rate of oxygen depletion in areas where oxygen-demanding wastes are present.

Words and Expressions

pollutant　[pə'l(j)u:t(ə)nt]　n. 污染物
discharge　[dis'tʃɑ:dʒ]　vt. vi. n. 排放
sewage　['su:idʒ]　n. 污水
urban　['ɜ:b(ə)n]　adj. 城市的
domestic　[də'mestik]　adj. 国内的；家庭的
pipe　[paip]　n. 管子
eliminate　[i'limineit]　vt. 消除；排除
treatment　['tri:tmənt]　n. 处理；治疗；对待
oxidize　['ɒksidaiz]　vt. vi. 使氧化；氧化
reduction　[ri'dʌkʃ(ə)n]　n. 减少；下降
consumption　[kən'sʌm(p)ʃ(ə)n]　n. 消耗；消费
dissolve　[di'zɒlv]　vt. vi. 使溶解（分解）；溶解（分解）
molecular　[mə'lekjʊlə]　adj. 分子的；由分子组成的
biodegradable　[,baiə(ʊ)di'greidəb(ə)l]　adj. 可生物降解的
organic　[ɔ:'gænik]　adj. 有机的；器官的；组织的
organism　['ɔ:g(ə)niz(ə)m]　n. 有机体；生物体；微生物
microorganism　[maikrəʊ'ɔ:g(ə)niz(ə)m]　n. 微生物；微小动植物
compound　['kɒmpaʊnd]　vt. 合成；混合
　　　　　　　　n. 化合物；混合物
　　　　　　　　adj. 混合的；复合的

depletion [di'pliːʃn] *n.* 消耗；损耗；耗尽
residue ['rezidjuː] *n.* 残渣；滤渣
nutrient ['njuːtriənt] *n.* 营养物质；营养盐；滋养物
nitrogen ['naitrədʒ(ə)n] *n.* 氮
phosphorus ['fɒsf(ə)rəs] *n.* 磷
excessive [ik'sesiv; ek-] *adj.* 过多的，极度的；过分的
proliferate [prə'lifəreit] *vt. vi.* 使激增；激增；增殖；扩散
algae [ˈældʒiː] *n.* 藻类；海藻
detergent [di'tɜːdʒ(ə)nt] *n.* 清洁剂；洗涤剂；洗衣粉
fertilizer ['fɜːtilaizə] *n.* 肥料
pathogenic [ˌpæθə'dʒenik] *adj.* 致病的；病原的；发病的（等于 pathogenetic）
bacteria [bæk'tiəriə] *n.* 细菌
viruses ['vaiərəsiz] *n.* 病毒；病霉（virus 的复数）
protozoa [ˌprəʊtə(ʊ)'zəʊə] *n.* [无脊椎] 原生动物；原生动物类（protozoan 的复数）
excrete [ik'skriːt; ek-] *vt.* 排泄；分泌
concentration [kɒns(ə)n'treiʃ(ə)n] *n.* 浓度；集中；浓缩；专心；集合
toxic ['tɒksik] *adj.* 有毒的；中毒的
toxicity [tɒk'sisəti] *n.* 毒性
particle ['pɑːtik(ə)l] *n.* 微粒；颗粒
sediment ['sedim(ə)nt] *n.* 沉积；沉淀物
settle ['set(ə)l] *vt.* 沉淀
turbidity [tɜː'bidəti] *n.* 浑浊；浑浊度
aquatic [ə'kwætik; -'kwɒt-] *adj.* 水生的；水栖的
pesticide ['pestisaid] *n.* 杀虫剂
herbicide ['hɜːbisaid] *n.* 除草剂
lead [liːd] *n.* 铅
zinc [ziŋk] *n.* 锌
exhaust [ig'zɔːst; eg-] *n.* 排气；废气；排气装置
　　　　　　　　　　　　vt. 排出；耗尽
　　　　　　　　　　　　vi. 排气
substance ['sʌbst(ə)ns] *n.* 物质
ecosystem ['iːkəʊsistəm] *n.* 生态系统
dispose [di'spəʊz] *vt. vi. n.* 处理；处置

surface water   地表水

domestic sewage   生活污水
municipal sewage   城市污水
point source pollution   点源污染
non-point source pollution   面源污染
agricultural runoff   农田径流，农田污水排放
natural drainage channel   自然排水沟
wastewater treatment   废水处理
prior to   在……之前；居先
oxygen-demanding material   耗氧物质；需氧物质
pose a threat to   对……造成威胁
critical level   临界水平；临界高度
suspended solid   悬浮物；悬浮体
colloidal particle   胶粒，胶体微粒
negative impact   负面影响
waste heat   废热

Notes

1. Domestic sewage and industrial wastes are called point sources because they are generally collected by a network of discharge into the receiving water of pipes or channels and conveyed to a single point of discharge into the receiving water. 生活污水和工业废水都称为水污染点源，这是因为它们通常都会被一个由各种管道或渠沟形成的网络收集起来，并集中到某个排放点排入收纳水体。
   - industrial waste   工业废弃物，课文此处指工业废水
   - point source   点源
   - non-point source   非点源
   - receiving water   收纳水体
   - convey   运输；传递；转达
   - be conveyed to   被传送到

     eg. I can't convey my feelings in words. 我无法用言语表达我的心情。

     In communications, the problem of electronics is how to convey information from one place to another. 在通讯系统中，电子设备要解决的问题是如何把信息从一个地方传递到另一个地方。

2. Often flow of polluted water flows over the surface of the land or along natural drainage channels to the nearest water body. 通常这类污水会在地表横流或者沿着天然的排水沟流入距离最近的水体。

- flow *n.* 流动；流量；泛滥；涨潮
    vt. 淹没；溢过
    vi. 流动；涌流
- flow over 溢出；横流
- drainage channel 排水沟；下水道
- water body 水体，水域，储水池

3. Anything that can be oxidized in the receiving water with the consumption of dissolved molecular oxygen is termed oxygen-demanding material. 所有在收纳水体中能够通过消耗水中的溶解氧而被氧化的物质都可以叫做耗氧物质。
    - that 从句作为定语从句修饰本句的主语 anything，关系代词 that 引导定语从句，并在定语从句中作主语，指导先行词 anything，不可省略；由于先行词 anything 是不定代词，that 不能替换成 which
    - term 术语；把……叫做
    - is termed=is called, is named 被命名为……；被称为……；被叫做……
        eg. Organic or inorganic particles that are carried by the wastewater into a receiving water are termed suspended solids (SS). 被废水带入某个收纳水体的有机微粒和无机微粒被称为悬浮固体。

4. As a rule, the most desirable commercial and game fish require high levels of dissolved oxygen. 通常最适合的商业鱼类和猎用鱼类的生存都需要高的溶解氧浓度。
    - game fish 供垂钓的鱼
    - as a rule 通常；一般来说

5. Nitrogen and phosphorus, two nutrients of primary concern, are considered pollutants because they are too much of a good thing. 氮和磷这两种最主要的营养物质也被看成污染物质，因为它们太多了也会变成坏事。
    - too much of a good thing 好事过头反成坏事；一件本来很好的事，一旦超过反而适得其反
        eg. Sunlight may be the best disinfectant, but there can be too much of a good thing. 阳光或许是最好的消毒剂，但凡事过犹不及。

6. Inorganic suspended solids are discharged by some industries but result mostly from soil erosion that is particularly bad in areas of logging, strip mining, and construction activity. 无机悬浮颗粒是从某些工厂企业排放出来，但多半是水土流失造成的，这在那些伐木、露天采矿和建筑施工活动区域情况尤其严重。
    - result from 起因于；由……造成
    - erode 侵蚀；腐蚀；冲刷
    - soil erosion *n.* 土壤侵蚀；水土流失

7. The lower James River in Virginia has been reduced to use only as a shipping channel because of a large industrial discharge of highly toxic and persistent organic compounds. 由于沿岸工厂产生的剧毒和持久性有机物的大量排放，弗吉尼亚州的詹姆士河下游现在仅能用作航运通道。
   - be reduced to 沦为；简化为；减小为；分解为
8. Higher temperatures also increase the rate of oxygen depletion in areas where oxygen-demanding wastes are present. 较高的水温也提高了存在耗氧废弃物的水域的耗氧率。
   - increase *vt. vi. n.* 增加；提高
     eg. The population continues to increase. 人口持续增长。
     It caused an increase of population in the area. 这导致了该地区的人口增长。

Exercises

I. Best choices.

1. The wide range of pollutants that are being discharged to _____ waters can be grouped into broad classes.
   A. rain　　　　B. ice　　　　C. fresh　　　　D. surface
2. Municipal sewage means both domestic sewage and _____ wastes.
   A. surface　　　B. urban　　　C. industrial　　　D. agricultural
3. Reduction of non-point source pollution is very _____.
   A. quick　　　　B. easy　　　　C. difficult　　　D. simple
4. DO means _____ in the third paragraph.
   A. defensive organization　　　B. dissolved oxygen
   C. depleted organism　　　　　D. domestic oxidization
5. _____ is not pathogenic organism.
   A. Bacteria　　　B. Viruses　　　C. Protozoa　　　D. Phosphorus

II. Decide whether each of the following statements is true (T) or false (F) according to the text.

(　) 1. Urban and agricultural runoff are called non-point sources.

(　) 2. Paper industry may produce oxygen-demanding wastes.

(　) 3. The abbreviation of organic or inorganic particles that are carried by the wastewater into a receiving water are termed BOD.

(　) 4. Small quantities of toxic compounds in the water can be compatible with the natural ecosystem and many human uses.

(　) 5. Heat is often recognized as a pollutant.

III. Translate the following words or phrases into English.

1. 污染物质 _____   2. 处理 _____

3. 排放 _____   4. 污水 _____

5. 微生物 _____   6. 营养物 _____

7. 藻类 _____   8. 微粒 _____

9. 废热 _____   10. 可生物降解的 _____

IV. Match the words and expressions with their meanings.

1. suspended solid   a. 有机化合物
2. domestic sewage   b. 耗氧物质
3. municipal sewage   c. 点源污染
4. DO   d. 病原菌
5. point source pollution   e. 生活污水
6. agricultural runoff   f. 工业废水
7. oxygen-demanding material   g. 溶解氧
8. industrial waste water   h. 农田径流
9. pathogenic organism   i. 城市污水
10. organic compound   j. 悬浮固体

V. Match the words and expressions with their meanings.

1. disposal   a. home
2. domestic   b. city
3. urban   c. treatment
4. increase   d. polluted water
5. consumption   e. reduce
6. sewage   f. depletion

VI. Translate the following sentences into Chinese.

1. Reduction of non-point source pollution generally requires changes in land use practices and improved education.

2. Some major sources of nutrients are phosphorus-based detergents, fertilizers, and food-processing wastes.

3. Certain shellfish can be toxic because they concentrate pathogenic organisms in their tissues, making the toxicity levels in the shellfish much greater than the levels in the surrounding water.

4. Even in rapidly moving mountain streams, sediment from mining and logging operations has destroyed many living places (ecological habitats) for aquatic organisms.

5. Many industrial wastewaters contain either toxic metals or toxic organic substances.

Unit 2　Primary Treatment

Text A　Wastewater Pretreatment (1)

The wide range of pollutants that are being discharged to surface waters can be grouped into broad classes. As civilization developed and cities grew, domestic sewage and industrial waste were eventually discharged into drainage ditches and sewers, and the entire contents emptied into the nearest watercourse. For major cities, this discharge was often enough to destroy even a large body of water.

Wastewater Characteristics

Discharges into a sanitary sewerage system consist of domestic wastewater, industrial discharge, and infiltration. The last adds to the total wastewater volume but is not itself a concern in wastewater disposal, infiltration will even dilute municipal sewage to some extent. These discharges vary widely with the size and type of industry and the amount of treatment applied before discharge into sewers.

Domestic sewage varies substantially in quantity and quality over time and from one community to the next. Typical variation for a small community is shown in Figure 1. Table 1 shows typical values for the most important parameters of domestic wastewater.

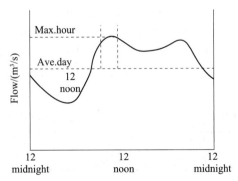

Figure 1　Daily variations in flow for a small community

Table 1 Characteristics of Typical Domestic Wastewater

Parameter	Typical value for domestic sewage
BOD/(mg/L)	250
SS/(mg/L)	220
Phosphorus/(mg/L)	8
Organic and ammonia nitrogen/(mg/L)	40
pH	6.8
Chemical oxygen demand/(mg/L)	500
total solids/(mg/L)	270

The treatment system selected to achieve these effluent standards includes:

(1)Primary treatment: physical processes that remove nonhomogenizable solids and homogenize the remaining effluent.

(2)Secondary treatment: biological processes that remove most of the biochemical demand for oxygen.

(3)Tertiary treatment: physical, biological, and chemical processes that remove nutrients like phosphorus, remove inorganic pollutants, deodorize and decolorize effluent water, and carry out further oxidation.

Screening and Grit Removal

The most objectionable aspect of discharging raw sewage into watercourses is the floating material. Thus screens were the first form of wastewater treatment used by communities, and are used today as the first step in treatment plants. Typical screens, shown in Figure 2, consist of a series of steel bars that might be about 2.5 cm apart. A screen in a modern treatment plant removes materials that might damage equipment or hinder further treatment. In some older treatment plants screens are cleaned by hand, but mechanical cleaning equipment is used in almost all new plants. The cleaning rakes are activated when screens get sufficiently clogged to raise the water level in front of the bars.

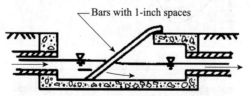

Figure 2　Bar screen used in wastewater treatment

In many plants, the second treatment step is a comminutor, a circular grinder designed to grind the solids coming through the screen into pieces about 0.3 cm or less in diameter. A typical comminutor design is shown in Figure3.

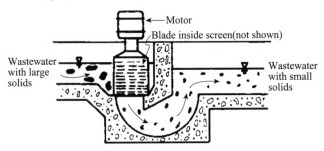

Figure 3   A comminutor used to grind up large solids

The third treatment step is the removal of grit or sand from the wastewater. Grit and sand can damage equipment like pumps and flow meters and, therefore, must be removed. The most common grit chamber is a wide place in the channel where the flow is slowed enough to allow the dense grit to settle out. Sand is about 2.5 times denser than most organic solids and thus settles much faster. The objective of a grit chamber is to remove sand and grit without removing organic materials. Organic materials must be treated further in the plant, but the separated sand may be used as fill without additional treatment.

Words and Expressions

primary   ['praim(ə)ri] *adj.* 主要的；初级的；基本的
civilization   [ˌsivilaiˈzeiʃən] *n.* 文明；文化
eventually   [i'ventʃuəli] *adv.* 最终
drainage   ['dreinidʒ] *n.* 排水
ditch   [ditʃ] *n.* 沟；沟渠
sewer   ['suːə; 'sjuːə] *n.* 下水道
watercourse   ['wɔːtəkɔːs] *n.* 河道；水道
characteristics   [ˌkærəktə'ristiks] *n.* 特性；特征
infiltration   [ˌinfil'treiʃən] *n.* 渗透物
volume   ['vɒljuːm] *n.* 容量
disposal   [di'spəuz(ə)l] *n.* 处理；清理
dilute   [dai'l(j)uːt; di-] *vt. vi.* 稀释；冲淡
vary   ['veəri] *vt. vi.* 变化；改变；使不同
variation   [veəri'eiʃ(ə)n] *n.* 变化

parameter [pə'ræmitə] n. 参数；系数
ammonia [ə'məuniə] n. [无化] 氨
effluent ['efluənt] n. 污水；流出物
homogenize [hə'mɒdʒənaiz] vt. vi. 使均匀；均化
physical ['fizik(ə)l] adj. 物理的
biological [baiə(ʊ)'lɒdʒik(ə)l] adj. 生物的
chemical ['kemik(ə)l] adj. 化学的
inorganic [inɔː'gænik] adj. 无机的；无生物的
deodorize [di'əudəraiz] vt. 除臭
decolorize [diː'kʌləraiz] vt. vi. 脱色
oxidation [ɒksi'deiʃ(ə)n] n. [化学] 氧化
screen [skriːn] n. 格栅
plant [plɑːnt] n. 工厂
hinder ['hində] vt. 阻碍
clog [klɒg] v. 阻塞；障碍
comminutor ['kɔminjuːtə] n. 破碎机
circular ['sɜːkjʊlə] adj. 圆形的
grind [graind] vt. vi. 研磨
diameter [dai'æmitə] n. 直径
grit [grit] n. 沙砾
pump [pʌmp] n. 泵
fill [fil] n. 填料；装填物

drainage ditch 排水沟
a body of water 水体
sewerage system 污水沟管系统
BOD （Biochemical Oxygen Demand）生化需氧量
COD （Chemical Oxygen Demand）化学需氧量
effluent standard 排污标准
primary treatment 一级处理；初级处理
secondary treatment 二级处理
tertiary treatment 高级/三级处理
raw sewage 未经处理的污水
floating material 悬浮物
steel bar 钢条

cleaning rake　清理耙
flow meter　流量计
grit chamber　沉砂池
settle out　沉积；沉淀出来
organic solid　有机颗粒

Notes

1. The last adds to the total wastewater volume but is not itself a concern in wastewater disposal, infiltration will even dilute municipal sewage to some extent. 其中渗透物本身并不需要处理，它仅增加了污水总量，甚至在一定程度上能稀释污水。
   - dilute 冲淡，稀释
     eg. Dilute the ammonia with water before you use it. 使用氨之前用水稀释它。
     　　The water will dilute the wine. 水能使酒变淡。
   - to some extent 在一定程度上；在某种程度上
     eg. So to some extent I'm for it. 所以在一定程度上，我支持它！
     　　To some extent you are correct. 在某种程度上你是正确的。

2. Domestic sewage varies substantially in quantity and quality over time and from one community to the next. 随着时间的推移，不同社区的生活污水的水质和水量有明显的变化。
   - 句子的主要成分是"Domestic sewage varies in quantity and quality"
   - over time and from one community to the next. 表示时间及地点的状语在翻译成中文时一般放在句首
   - substantially 本质上；明显地
     eg. The price may go up quite substantially. 价格可能大大上扬。

3. A screen in a modern treatment plant removes materials that might damage equipment or hinder further treatment. 现代污水处理厂的格栅是用于去除污水中较大的悬浮物，这些物质可能会损坏设备或阻碍进一步的处理。
   - 句子的主要成分是"A screen removes materials"
   - 主语 screen 受到介词短语 in a modern treatment plant 的修饰，翻译成中文时一般放在所修饰词的前面
   - 宾语 materials 受到 that 从句修饰。鉴于定语从句比较长，不便于将它译成汉语中的前置定语的形式，故宜把它拆开另作一句
   - hinder 阻碍；妨碍
     eg. Don't hinder me in my work. 不要妨碍我的工作。

4. The cleaning rakes are activated when screens get sufficiently clogged to raise the water level in front of the bars. 当格栅被堵塞，导致钢条前的水位提升时，清理耙就会启动。
   - 句子的主要成分是"The cleaning rakes are activated"
   - activate 启动；激活
     eg. These push buttons can activate the elevator. 这些按钮能启动电梯。
         We must activate the youth to study. 我们要激励青年去学习。
   - clog 堵塞；阻碍
     eg. In cotton and wool processing, short length fibers may clog sewers. 在棉毛生产中，短纤维可能堵塞下水管道。
5. Grit and sand can damage equipment like pumps and flow meters and, therefore, must be removed. 沙砾会损坏如水泵和流量计等设备，因此必须被去除。
   - like 比如；例如
     eg. They export a lot of fruit, like apples, oranges, lemons, etc. 他们进口许多水果，比如苹果、柑橘和柠檬等。
   - and 连接两个并列谓语，并用"therefore"表因果关系
     eg. Simply that willpower is the source of our action, and, therefore, the source of our character. 很简单，意志力是我们行动的源泉，因此也是我们品质的源泉。

Exercises

I. Answer the following questions after reading the text.

1. Why do we have to treat the domestic sewage and industrial wastewater?
2. What is the objective of primary treatment?
3. Generally speaking, what is the sewage flow's variation characteristic of a small community?
4 Why should the bar screen be the first step in treatment plants?
5. What is the objective of a grit chamber?

II. Decide whether each of the following statements is true (T) or false (F) according to the text.

1. Infiltration itself is a concern in wastewater disposal.
2. Primary treatment is biological process that remove most of the biochemical demand for oxygen.
3. In almost all new plants screens are cleaned by hand.
4. A screen can remove suspended solids.

5. Comminutor is generally used in primary treatment, belonging to the physical processes.

III. Translate the following words or phrases into English.

1. 下水道 _____

2. 初级处理 _____

3. 清理耙 _____

4. 泵 _____

5. 未经处理的污水 _____

6. 格栅 _____

7. 氨 _____

8. 悬浮物 _____

9. 破碎机 _____

10. 化学需氧量 _____

IV. Translate the following sentences into Chinese.

1. Discharges into a sanitary sewerage system consist of domestic wastewater, industrial discharge, and infiltration.

2. Primary treatment: physical processes that remove nonhomogenizable solids and homogenize the remaining effluent.

3. Thus screens were the first form of wastewater treatment used by communities, and are used today as the first step in treatment plants.

4. In many plants, the second treatment step is a comminutor, a circular grinder designed to grind the solids coming through the screen into pieces about 0.3 cm or less in diameter.

5. The most common grit chamber is a wide place in the channel where the flow is slowed enough to allow the dense grit to settle out.

V. Match the words and expressions with their meanings.

( ) 1. organic solid   a. 除臭

( ) 2. BOD   b. 排水沟

( ) 3. drainage ditch   c. 特性

( ) 4. deodorize   d. 生化需氧量
( ) 5. flow meter   e. 二级污水处理
( ) 6. characteristics   f. 有机颗粒
( ) 7. grit chamber   g. 排污标准
( ) 8. phosphorus   h. 沉砂池
( ) 9. secondary treatment   i. 流量计
( ) 10. effluent standard   j. 磷

VI. Best choices.

1. Discharges into a sanitary sewerage system consist of _____.
   A. domestic wastewater   B. industrial discharge
   C. infiltration   D. all of the above

2. Primary treatment is _____ that remove nonhomogenizable solids and homogenize the remaining effluent.
   A. physical process   B. biological process
   C. chemical process   D. all of the above

3. What is the first step used in wastewater treatment plants nowadays?
   A. screen   B. grit chamber   C. settling tank   D. comminutor

4. _____ are activated when screens get sufficiently clogged to raise the water level in front of the bars.
   A. The steel bars   B. The cleaning rakes
   C. The flow meter   D. all of the above

5. What can a screen remove?
   A. Sludge.   B. Oil and grease.   C. Suspended solids.   D. all of the above.

VII. Fill in the following blanks in the figure.

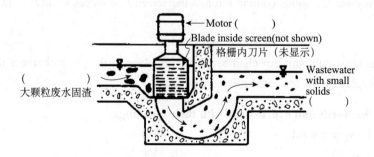

Text B Wastewater Pretreatment (2)

Settling Tank

Most wastewater treatment plants have a settling tank after the grit chamber, to settle out as much solid materials as possible. Accordingly, the retention time is long and turbulence is kept to a minimum. Retention time is the total time an average slug of water spends in the tank and is calculated as the time required to fill the tank. For example, if the tank volume is 100 m^3 and the flow rate is 2 m^3/min, the retention time is 100/2 = 50 min. Generally, two types of sedimentation basins (also called tanks, or clarifiers) are used: rectangular(Figure 1) and circular(Figure 2).

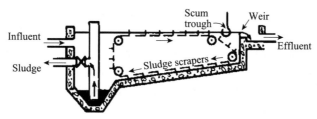

Figure 1 Rectangular settling tank

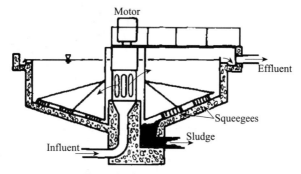

Figure 2 Circular settling tank

The solids settle to the bottom of the tank and are removed through a pipe, while the clarified liquid escapes over a V-notch weir, which distributes the liquid discharge equally all the way around a tank. Settling tanks are also called sedimentation tanks or clarifiers. The settling tank that immediately follows screening and grit removal is called the primary clarifier. The solids that drop to the bottom of a primary clarifier are removed as raw sludge.

Raw sludge generally has a powerfully unpleasant odor and is full of water, two characteristics that make its disposal difficult. It must be stabilized to reduce further decomposition and dewatered for ease of disposal. Solids from processes other than the primary clarifier must be treated similarly before disposal.

Flotation

Flotation is an operation that removes not only oil and grease but also suspended solids. It is discussed in this section since it is one of the most effective systems for suspensions which contain oil and grease. The most common procedure is that of dissolved air flotation (DAF), in which the waste stream is first pressurized with air in a closed tank. After passing through a pressure-reduction valve, the wastewater enters the flotation tank (Figure 3) where, due to the sudden reduction in pressure, minute air bubbles in the order of 50~100 microns in diameter are formed. As the bubbles rise to the surface, the suspended solids and oil or grease particles adhere to them and are carried upwards. It is common practice to use chemicals to enhance flotation performance. As with coagulants (discussed later) these aids should preferably be innocuous, since these recovered solids are frequently used in animal feed formulations.

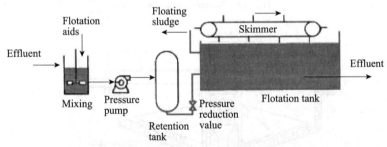

Figure 3　Diagram of a dissolved air flotation system

One alternate design involves the recycling of part (10%~30%) of the treated water (Figure 4). All systems contain a mechanism for removing the solids that may settle to the bottom of the flotation tanks, usually by a helical conveyor placed in the conical bottom. The main advantage claimed of DAF systems is the faster rate at which very small or light suspended solids can be removed in comparison with settling.

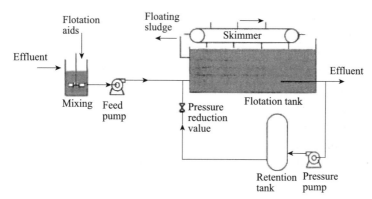

Figure 4　Diagram of a DAF system with recycle

Performance of DAF systems has been reported to be dependent on several factors, of which one of the most important is the solids concentration; higher solids content usually gives higher removal efficiencies. Other factors affecting the efficiency of the operation are the ratio of air to solids (A/S) defined as the amount of air released after pressure reduction and the amount of solids present in the wastewater. There is usually an optimum A/S which is determined by bench scale tests.

Key factors in the successful operation of DAF units are the maintenance of proper pH (usually between 4.5 and 6, with 5 being most common to minimize protein solubility and break-up emulsions), proper flow rates and the continuous presence of trained operators.

Words and Expressions

turbulence　['tɜ:bjʊl(ə)ns] *n.* 湍流；涡流；搅动
minimum　['miniməm] *n.* 最小值；最低限度
calculate　['kælkjʊleit] *vt. vi.* 计算
sedimentation　[,sedimən'teiʃən] *n.* [化学] 沉淀
basin　['beis(ə)n] *n.* 水池
clarifier　['klærifaiə] *n.* 澄清池
rectangular　[rek'tæŋgjʊlə] *adj.* 矩形的
weir　[wiə] *n.* 堰
sludge　['slʌdʒ] *n.* 污泥
stabilize　['steibəlaiz] *vt. vi.* 稳定；稳固
decomposition　[,di:kɒmpə'ziʃn] *n.* 分解，腐烂
dewater　[di:'wɔ:tə] *vt.* 使脱水

flotation [fləʊ'teiʃ(ə)n] n. 气浮
grease [griːs] n. 油脂
suspension [sə'spenʃ(ə)n] n. 悬浮物
procedure [prə'siːdʒə] n. 程序；步骤
pressurize ['preʃəraiz] vt. 增压；使……加压
valve [vælv] n. 阀
minute ['minit] adj. 微小的
bubble ['bʌb(ə)l] n. 气泡，泡沫
micron ['maikrɒn] n. 微米（等于百万分之一米）
adhere [əd'hiə] vt. vi. 黏附
coagulant [kəʊ'ægjʊlənt] n. 混凝剂
innocuous [i'nɒkjʊəs] adj. 无害的
formulation [fɔːmjʊ'leiʃn] n. 配方；配制；制剂
recycling [riː'saikliŋ] n.（资源、垃圾的）回收利用
mechanism ['mek(ə)niz(ə)m] n. 机械装置
factor ['fæktə] n. 因素；要素
efficiency [i'fiʃ(ə)nsi] n. 效率；效能
optimum ['ɒptiməm] n. 最佳效果
maintenance ['meint(ə)nəns; -tin-] n. 维护；维修；保持
solubility [ˌsɒljʊ'biləti] n. 溶解度
emulsion [i'mʌlʃ(ə)n] n. 乳液

settling tank　沉淀池
retention time　停留时间
flow rate　流速
sedimentation basin　沉淀池
scum trough　浮渣去除槽
sludge scraper　刮渣机
clarified liquid　澄清的液体
V-notch weir　三角堰
primary clarifier　初级澄清池
raw sludge　初次沉淀污泥
DAF （Dissolved Air Flotation）溶气气浮
closed tank　密封罐
pressure-reduction valve　泄压阀
flotation tank　气浮池
air bubble　气泡

grease particle　油脂粒
animal feed formulations　动物饲料配方
flotation aids　浮选剂
pressure pump　加压泵
retention tank　溶气罐
floating sludge　浮渣
helical conveyor　螺旋输送机
conical bottom　锥形底
solids concentration　固体颗粒浓度
removal efficiencies　去除率
bench scale test　实验室规模测试
protein solubility　蛋白质溶解度

Notes

1. The settling tank that immediately follows screening and grit removal is called the primary clarifier. 沉淀池与格栅、沉砂池一起被称为初级澄清池。
   - 句子的主要成分是"The settling tank is called the primary clarifier"
   - that 引导定语从句修饰主语 the settling tank（沉淀池）
   - grit removal 沉砂池

2. The solids that drop to the bottom of a primary clarifier are removed as raw sludge. 沉入初级澄清池并被排出的固体被称为初次沉淀污泥。
   - 句子的主要成分是"The solids are removed as raw sludge"
   - that 引导定语从句修饰主语 the solids（固体）
   - remove 去除；取出
     eg. Filters do not remove all contaminants from water. 过滤器无法过滤掉水中的所有污染物。

3. Flotation is an operation that removes not only oil and grease but also suspended solids. 气浮是从液体中去除油脂和悬浮颗粒的一种方法。
   - 句子的主要成分是"Flotation is an operation"
   - that 引导定语从句修饰 operation（方法）
   - oil and grease 油脂
   - suspended solids 悬浮颗粒

4. As with coagulants (discussed later) these aids should preferably be innocuous, since these recovered solids are frequently used in animal feed formulations. 与混凝剂一样，浮选剂应该是无害的，因为回收的固体经常用作动物饲料配方。
   - as with 与……一样；正如

eg. As with any meeting, you want to know your participants well. 与任何会议一样，您需要充分了解参会者。
- since 因为；由于

  eg. I'm forever on a diet, since I put on weight easily. 我永远都在减肥，因为我很容易长胖。

  Since she did not make enough money to live in her own house, she went back to live with her mother. 她挣的钱不够自己一个人住，于是搬回去和她妈妈一起住了。
- animal feed formulations 动物饲料配方

5. The main advantage claimed of DAF systems is the faster rate at which very small or light suspended solids can be removed in comparison with settling. 与沉淀池相比，溶气气浮的主要优点是细小悬浮物可以更快被去除。
- 句子的主要成分是："The advantage of DAF systems is the faster rate"
- at which 引导的限定性定语从句用来修饰宾语 faster rate
- in comparison with 与……相比

  eg. That old building is not so imposing as it used to be in comparison with other tall odern structures. 和其他的高大现代化建筑相比，那座古建筑没有过去那么壮观了。

Exercises

I. Answer the following questions after reading the text.

1. What is the definition of retention time?
2. Why do we need the V-notch weir on the reactor?
3. How do the DAF systems remove the suspended solids and oil or grease?
4. Which factor is the most important that DAF systems performance depend on?
5. How can we get the optimum A/S data?

II. Decide whether each of the following statements is true (T) or false (F) according to the text.

( ) 1. Sedimentation basins usually have two types: Rectangular and Circular.

( ) 2. Raw sludge's disposal is very convenient.

( ) 3. It is common practice to use chemicals to enhance flotation performance.

( ) 4. As the bubbles rise to the surface, the suspended solids settle down to the bottom.

( ) 5. Proper pH is one of the key factors in the successful operation of DAF units.

III. Translate the following words or phrases into English.

1. 初级澄清池 _____
2. 污泥 _____
3. 三角堰 _____
4. 气浮 _____
5. 加压泵 _____
6. 配方 _____
7. 混凝剂 _____
8. 浓度 _____
9. 沉淀 _____
10. 去除率 _____

IV. Translate the following sentences into Chinese.

1. Retention time is the total time an average slug of water spends in the tank and is calculated as the time required to fill the tank.

2. The solids settle to the bottom of the tank and are removed through a pipe, while the clarified liquid escapes over a V-notch weir, which distributes the liquid discharge equally all the way around a tank.

3. As the bubbles rise to the surface, the suspended solids and oil or grease particles adhere to them and are carried upwards.

4. All systems contain a mechanism for removing the solids that may settle to the bottom of the flotation tanks, usually by a helical conveyor placed in the conical bottom.

5. Performance of DAF systems has been reported to be dependent on several factors, of which one of the most important is the solids concentration; higher solids content usually gives higher removal efficiencies.

V. Match the words and expressions with their meanings.

( ) 1. retention time   a. 气浮池
( ) 2. DAF   b. 刮渣机
( ) 3. sludge scraper   c. 停留时间
( ) 4. flotation tank   d. 油脂粒
( ) 5. settling tank   e. 蛋白质溶解度
( ) 6. grease particle   f. 气泡
( ) 7. pressure-reduction valve   g. 螺旋输送机
( ) 8. protein solubility   h. 泄压阀

( ) 9. air bubble　　　　　　　i. 沉淀池
( ) 10. helical conveyor　　　　j. 溶气气浮

VI. Best choices.

1. If the tank volume is 120 m^3 and the flow rate is 2 m^3/min, how long is the retention time?
 A. 30 min　　　B. 40 min　　　C. 50 min　　　D. 60 min
2. What is called the primary clarifier?
 A. The settling tank　B. Screening　　C. Grit removal　D. all of the above
3. Flotation is an operation that removes not only oil and grease but also _____.
 A. suspended solids　B. grease particle　C. protein　　D. phosphorus
4. As with coagulants, these aids should preferably be _____.
 A. nocuous　　　B. innocuous　　C. favorably　　D. feasible
5. What are the Key factors in the successful operation of DAF units?
 A. The maintenance of proper pH　　　　　　　B. Proper flow rates
 C. The continuous presence of trained operators　D. all of the above

VII. Fill in the following blanks in the figures.

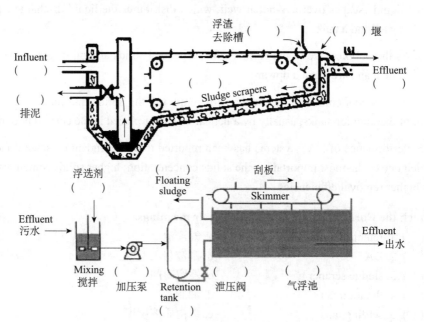

Unit 3 Secondary Treatment

Text A Aerobic Treatment-Activated Sludge System

Water leaving the primary clarifier has lost much of the solid organic matter but still contains high-energy molecules that decompose by microbial action, creating BOD. The demand for oxygen must be reduced (energy wasted) or else the discharge may create unacceptable conditions in the receiving waters. The objective of secondary treatment is to remove BOD, whereas the objective of primary treatment is to remove solids.

An activated sludge system, as shown in the block diagram in the figure, includes a tank full of waste liquid from the primary clarifier and a mass of microorganisms. Air bubbled into this aeration tank provides the necessary oxygen for survival of the aerobic organisms. The microorganisms come in contact with dissolved organic matter in the wastewater, adsorb this material, and ultimately decompose the organic material to CO_2, H_2O, some stable compounds, and more microorganisms. The production of new organisms is relatively slow, and uses most of the aeration tank volume.

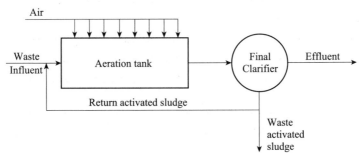

Figure1 Block diagram of an activated sludge system

When most of the organic material, which is food for the microorganisms, has been used up, the microorganisms are separated from the liquid in a settling tank, sometimes called a secondary or final clarifier. The microorganisms remaining in the settling tank have no food available, become hungry, and are thus activated; hence the term activated sludge. The clarified liquid escapes over a weir and may be discharged into the receiving water. The settled microorganisms, now called return activated sludge, are pumped back to the head of the aeration tank where they find more food in the organic compounds in the liquid entering the aeration tank from the primary clarifier, and the process starts over again. Activated sludge treatment is a continuous process, with continuous sludge pumping and clean water discharge.

Activated sludge treatment produces more microorganisms than necessary and if the microorganisms are not removed, their concentration will soon increase and clog the system with solids. Some of the microorganisms must therefore be wasted. Disposal of such waste activated sludge is one of the most difficult aspects of wastewater treatment.

Activated sludge systems are designed on the basis of loading, or the amount of organic matter, or food, added relative to the microorganisms available. The food-to-microorganism (F/M) ratio is a major design parameter. Both F and M are difficult to measure accurately, but may be approximated by influent BOD and SS in the aeration tank, respectively. The combination of liquid and microorganisms undergoing aeration is known as mixed liquor, and the SS in the aeration tank are mixed liquor suspended solids (MLSS). The ratio of influent BOD to MLSS, the F/M ratio, is the loading on the system, calculated as pounds (or kg) of BOD per day per pound or kilogram of MLSS.

Relatively small F/M, or little food for many microorganisms, and a long aeration period (retention time in the tank) result in a high degree of treatment because the microorganisms can make maximum use of available food. Systems with these features are called extended aeration systems and are widely used to treat isolated wastewater sources, like small developments or resort hotels. Extended aeration systems create little excess biomass and little excess activated sludge to dispose of.

Secondary treatment of wastewater usually includes a biological step like activated sludge, which removes a substantial part of the BOD and the remaining solids. The typical wastewater that we began with now has the following approximate water quality.

Table1 Approximate water quality

	Raw wastewater	After primary treatment	After secondary treatment
BOD (mg/L)	250	175	15
SS (mg/L)	220	60	15
P (mg/L)	8	7	6

The effluent from secondary treatment meets the previously established effluent standards for BOD and SS. Only phosphorus content remains high. The removal of inorganic compounds, including inorganic phosphorus and nitrogen compounds, requires advanced or tertiary wastewater treatment.

Words and Expressions

  clarifier ['klærifaiə] *n.* 澄清器
  solid ['sɒlid] *adj.* 固体的；可靠的；立体的；结实的；一致的
  molecule ['mɒlikjuːl] *n.* 分子；微小颗粒，微粒
  decompose [diːkəm'pəʊz] *vi. vt.*（使）分解；（使）腐烂
  microbial [mai'krəʊbiəl] *adj.* 微生物的；由细菌引起的
  oxygen ['ɒksidʒ(ə)n] *n.* 氧气；氧
  activated ['æktiveitid] *adj.* 活性化的；活泼的
  diagram ['daiəgræm] *n.* 图表；图解
  tank [tæŋk] *n.* 水槽
  aeration [eiə'reiʃən] *n.* [环境] 曝气
  survival [sə'vaiv(ə)l] *n.* 生存
  aerobic [eə'rəʊbik] *n.* 需氧的
  adsorb [əd'zɔːb; -'sɔːb] *vt.* 吸附；吸收
  ultimately ['ʌltimətli] *adv.* 最后
  compound ['kɒmpaʊnd] *n.* [化学] 化合物
  volume ['vɒljuːm] *n.* 容积
  available [ə'veiləb(ə)l] *adj.* 可获得的；可利用的
  weir [wiə] *n.* 出水堰
  pump [pʌmp] *vi. vt.* 抽水
  clog [klɒg] *v.* 阻塞；障碍
  ratio ['reiʃiəʊ] *n.* 比率；比例
  accurately ['ækjərətli] *adv.* 精确地，准确地
  approximate [ə'prɒksimət] *vt.* 粗略估计
  influent ['influənt] *adj.* 流入的
  undergo [ʌndə'gəʊ] *vt.* 经历
  calculate ['kælkjʊleit] *vt.* 计算
  retention [ri'tenʃ(ə)n] *n.* 保留
  maximum ['mæksiməm] *adj.* 最高的；最多的；最大极限的
  biomass ['baiə(ʊ)mæs] *n.* 生物量

biological [baiə(ʊ)'lɒdʒik(ə)l] *adj.* 生物的
substantial [səb'stænʃ(ə)l] *adj.* 大量的
inorganic [inɔː'gænik] *adj.* 无机的；无生物的
tertiary ['tɜːʃ(ə)ri] *adj.* 第三的

primary clarifier   初级澄清池
organic matter   有机物；有机物质
activated sludge   活性污泥
block diagram   框图；方块图
aeration tank   曝气池
aerobic organism   好氧微生物
secondary / final clarifier   二级 / 最终澄清池
clarified liquid   上清液
return activated sludge   回流活性污泥
organic compound   有机化合物
start over again   从头开始，又一次开始
on the basis of   基于，根据
design parameter   设计参数
SS   悬浮固体浓度
MLSS   混合液悬浮固体浓度
retention time   停留时间
extended aeration system   延时曝气系统
phosphorus content   磷含量
inorganic compound   无机（化合）物
inorganic phosphorus   无机磷
nitrogen compound   含氮化合物
advanced/ tertiary wastewater treatment   高级 / 三级废水处理

Notes

1. Water leaving the primary clarifier has lost much of the solid organic matter but still contains high-energy molecules that decompose by microbial action, creating BOD. 由初级澄清池排出的水已经去除大部分的有机物，但是仍然含有一些高分子有机物，在微生物的作用下，高分子有机物分解会产生 BOD。
    - high-energy molecules 高分子
    - ...that decompose by microbial action that 引导定语从句

- ...creating BOD 现在分词作伴随状语，表示高分子有机物在分解的同时会产生BOD。

2. An activated sludge system, as shown in the block diagram in the figure, includes a tank full of waste liquid from the primary clarifier and a mass of microorganisms. 如图所示，一个活性污泥系统，包括一个装满了来自初级澄清池的废水和大量微生物的水池。

   - as shown in 如……所示
     eg. By using the two together, you can copy from one directory to another, as shown in Listing 13. 同时使用这两个选项，你可以将文件从一个目录复制到另一个目录，如清单13所示。
   - full of 装满，尽是
     eg. His head is full of nonsense. 他满脑子荒唐念头。
     The meeting was full of friendly atmosphere from beginning to end. 会议始终充满友好的气氛。
   - a mass of 大量的

3. The microorganisms come in contact with dissolved organic matter in the wastewater, adsorb this material, and ultimately decompose the organic material to CO_2, H_2O, some stable compounds, and more microorganisms. 微生物接触废水中的溶解性有机物，吸附它们，并最终将这些有机物分解为二氧化碳、水、一些稳定的化合物以及新生繁殖的微生物。

   - come in contact with 接触
     eg. If your hands come in contact with germs in a dirty bathroom and you touch your eyes, nose, or mouth, the bacteria can enter our body and cause diseases. 如果你的手接触到脏卫生间里的细菌，然后又接触到你的眼睛、鼻子或者是嘴巴，这种细菌就能够侵入你的身体，从而导致疾病。
   - dissolved organic matter 溶解性有机物

4. When most of the organic material, which is food for the microorganisms, has been used up, the microorganisms are separated from the liquid in a settling tank, sometimes called a secondary or final clarifier. 当作为微生物食物的大部分有机物被消耗殆尽时，微生物便在沉淀池与液体分离（沉淀池有时也被称为二级或最终澄清池）。

   - use up 用完，耗尽
     eg. By this time he had used up all his savings. 到这时，他的存款已全部用完。
   - be separated from 与……分离
     eg. Separate the whites from the yolks. 将蛋清和蛋黄分开。

5. The settled microorganisms, now called return activated sludge, are pumped back to the head of the aeration tank where they find more food in the organic compounds in the

liquid entering the aeration tank from the primary clarifier, and the process starts over again. 遗留下的微生物（现可称为回流活性污泥）由泵送回曝气池前端，在曝气池前端有由初级澄清池注入曝气池的污水，微生物们可以在污水中的有机化合物里觅到更多的食物，此过程一遍又一遍循环反复。

- return activated sludge 回流活性污泥
- organic compounds 有机化合物
- start over again 从头开始，又一次开始

  eg. I didn't go well. May I start over again? 我没弄好，能重新再来一遍吗？

6. Relatively small F/M, or little food for many microorganisms, and a long aeration period (retention time in the tank) result in a high degree of treatment because the microorganisms can make maximum use of available food. 相对较小的 F/M 比例，或少量食物对应大量微生物，以及长时间的曝气周期（曝气池中的停留时间）都可引起高负荷处理，其原因在于微生物能够最大限度地利用可用食物。

- retention time 停留时间
- make use of 使用，利用

  eg. They make use of advertisements to plug the new product. 他们利用广告宣传这种新产品。

7. The effluent from secondary treatment meets the previously established effluent standards for BOD and SS. Only phosphorus content remains high. The removal of inorganic compounds, including inorganic phosphorus and nitrogen compounds, requires advanced or tertiary wastewater treatment. 二级处理后的废水符合国内目前 BOD 和 SS 排放标准，但是磷含量仍然超标。如要去除包括无机磷和含氮化合物在内的无机化合物，则需要高级或三级废水处理。

- effluent standard 排污标准
- advanced or tertiary wastewater treatment 高级或三级废水处理

Exercises

I. Answer the following questions after reading the text.

    1. What is the objective of primary treatment, and what about that of the secondary treatment?

    2. What is the purpose of air bubbles in an aeration tank?

    3. How is activated sludge produced?

    4. Why should we dispose waste activated sludge?

    5. How can we get F and M?

II. Decide whether each of the following statements is true (T) or false (F) according to the text.

(　) 1. Decomposing microorganisms will create BOD.

(　) 2. An activated sludge system includes a tank full of waste liquid from the secondary clarifier and a mass of microorganisms.

(　) 3. The clarified liquid escapes over a weir and may be discharged directly into the receiving water.

(　) 4. The settled microorganisms, now called return activated sludge, are pumped back to the head of the settling tank.

(　) 5. The effluent from secondary treatment meets the previously established effluent standards only for BOD.

III. Translate the following words or phrases into English.

1. 初级澄清池 _____　　2. 初级处理 _____

3. 活性污泥 _____　　4. 好氧微生物 _____

5. 最终澄清池 _____　　6. 收纳水体 _____

7. 回流活性污泥 _____　　8. 延时曝气系统 _____

9. 二级处理 _____　　10. 排污标准 _____

IV. Translate the following sentences into Chinese.

1. The objective of secondary treatment is to remove BOD, whereas the objective of primary treatment is to remove solids.

2. Air bubbled into this aeration tank provides the necessary oxygen for survival of the aerobic organisms.

3. The microorganisms remaining in the settling tank have no food available, become hungry, and are thus activated; hence the term activated sludge.

4. Activated sludge treatment produces more microorganisms than necessary and if the microorganisms are not removed, their concentration will soon increase and clog the system

with solids.

5. The combination of liquid and microorganisms undergoing aeration is known as mixed liquor, and the SS in the aeration tank are mixed liquor suspended solids (MLSS).

V. Match the words and expressions with their meanings.

( ) 1. organic matter　　　　　　a. 悬浮固体浓度
( ) 2. BOD　　　　　　　　　　b. 曝气池
( ) 3. block diagram　　　　　　c. 混合液悬浮固体浓度
( ) 4. aeration tank　　　　　　　d. 生化需氧量
( ) 5. settling tank　　　　　　　e. 三级废水处理
( ) 6. SS　　　　　　　　　　　f. 有机物；有机物质
( ) 7. MLSS　　　　　　　　　　g. 框图；方块图
( ) 8. phosphorus content　　　　h. 设计参数
( ) 9. tertiary wastewater treatment　i. 沉淀池
( ) 10. design parameter　　　　　j. 磷含量

VI. Best choices.

1. Microorganisms decompose organic materials to _____.
   A. CO_2, H_2O　　　　　　B. some stable compounds
   C. more microorganisms　　　D. all of the above

2. Microorganisms are separated from the liquid in _____.
   A. a primary clarifier　　　　　B. an aeration tank
   C. a final clarifier　　　　　　D. a digesting tank

3. What are the foundations of designing activated sludge systems?
   A. loading
   B. the amount of organic matter
   C. food added relative to the microorganisms available
   D. all of the above

4. What are the features of an extended aeration system?
   A. small F/M　　　　　　　　B. little food for many microorganisms
   C. a long aeration period　　　D. all of the above

5. Activated sludge is a _____ step.
   A. biological　　B. chemica　　C. physical　　D. medical

VII. Fill in the following blanks in the figure.

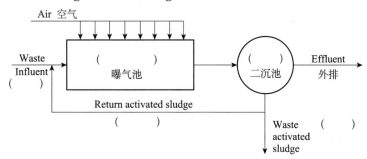

Text B Anaerobic Wastewater Treatment

Anaerobic wastewater treatment uses biological agents in an oxygen-free environment to remove impurities from wastewater. After undergoing such a treatment, water can be safely released back into the environment. The biological agents used in the process are microorganisms that consume or break down biodegradable materials in sludge, or the solid portion of wastewater following its filtration from polluted water.

Anaerobic wastewater treatment is also known as anaerobic digestion due to the action of the microorganisms. That is, they are essentially "digesting" the polluted parts of the water. An excellent way to decrease the amount of organic matter leftover in things such as sewage and leftover food, anaerobic digestion is typically a component of any biological wastewater treatment system.

An anaerobic treatment system is a complex three-step process that produces methane gas (in addition to other products) from the biological digestion of sewage waste. The first stage is the hydrolysis of lipids, cellulose, and protein. Extracellular enzymes produced by the inhabiting bacteria break down these macromolecules into smaller and more digestible forms. Next, these molecules are decomposed into fatty acids. This decomposition is performed by several facultative and anaerobic bacteria. Finally, methanogenic bacteria digest these fatty acids, resulting in the formation of methane gas.

Usually, the anaerobic process takes place in sealed tanks, located either above or below the ground. During the initial stages of the sludge breakdown, the microorganisms, which are mostly bacteria, convert the waste into organic acids, ammonia, hydrogen and carbon dioxide. In the final stages of anaerobic wastewater treatment, the remains of the sludge are converted, by a single-celled microorganism known as a methanogen, into a biogas consisting of methane and carbon dioxide.

An additional benefit of anaerobic wastewater treatment is its reduction of gas emissions. The biogas that results from the anaerobic wastewater treatment may actually be harnessed and used as an alternative power source for cooking, lighting, heating and engine fuel. In other words, by capturing and utilizing the methane and carbon dioxide produced by anaerobic digestion, the biogas is not released into the atmosphere.

Many in the scientific community believe that high concentrations of methane and carbon dioxide, also known as greenhouse gases, in Earth's atmosphere, contribute to the process of global warming. This theory, known as the greenhouse effect, postulates that these gases trap heat from the sun in the atmosphere, thereby increasing global temperature. While the theory has led to some controversy, using biogas as an alternative to fossil fuels has some practical applications.

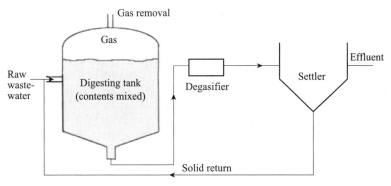

Figure 1 Diagram of an anaerobic digestion process

Words and Expressions

anaerobic   [ˌæneəˈrəʊbik] *adj.* 厌氧的
agent   [ˈeidʒ(ə)nt] *n.* 药剂
impurity   [imˈpjʊəriti] *n.* 杂质
consume   [kənˈsjuːm] *vt.* 消耗
polluted   [pəˈljuːtid] *adj.* 受污染的
filtration   [filˈtreiʃn] *n.* 过滤
methane   [ˈmiːθein; ˈmeθein] *n.* 甲烷
digestion   [daiˈdʒestʃ(ə)n; di-] *n.* 消化
essentially   [iˈsenʃ(ə)li] *adv.* 本质上；本来
decrease   [diˈkriːs] *vt.* 减少；减小
leftover   [ˈleftəʊvə] *n.* 剩余物；残留物
             adj. 残余的
component   [kəmˈpəʊnənt] *n.* 成分；组成
hydrolysis   [haiˈdrɒlisis] *n.* 水解作用
lipid   [ˈlipid] *n.* 脂肪；脂质；油脂
cellulose   [ˈseljʊləʊz; -s] *n.* 纤维素
protein   [ˈprəʊtiːn] *n.* 蛋白质
extracellular   [ekstrəˈseljʊlə] *adj.*（位于或发生于）细胞外的
enzyme   [ˈenzaim] *n.* 酶
inhabit   [inˈhæbit] *vi. vt.* 栖息；居住于

macromolecule [ˌmækrə(ʊ)'mɒlikjuːl] n. 大分子
digestible [dai'dʒestəbl] adj. 易消化的
fatty ['fæti] adj. 脂肪的
acid ['æsid] n. 酸
decomposition [ˌdiːkɒmpə'ziʃn] n. 分解
facultative ['fæk(ə)l,tətiv] n. 兼性
methanogenic [ˌmeθənə'dʒenik] adj. 产甲烷的
sealed [siːld] adj. 密封的
initial [i'niʃəl] adj. 最初的
convert [kən'vɜːt] vi. vt. (使)转变；转换
ammonia [ə'məʊniə] n. 氨
hydrogen ['haidrədʒ(ə)n] n. 氢
carbon ['kɑːb(ə)n] n. 碳
 adj. 碳的
dioxide [dai'ɒksaid] n. 二氧化物
methanogen [mə'θænədʒən] n. 产烷生物
reduction [ri'dʌkʃn] n. 减少；下降；缩小
emission [i'miʃ(ə)n] n. 排放
harness ['hɑːnis] vt. 治理；利用
alternative [ɔːl'tɜːnətiv; ɒl-] adj. 供选择的；选择性的；交替的
 n. 二中择一；供替代的选择
atmosphere ['ætməsfiə] n. 大气；空气；大气层
postulate ['pɒstjʊleit] vt. 假定；要求；视……为理所当然
controversy ['kɒntrəvɜːsi; kən'trɒvəsi] n. 争论；论战；辩论
fossil ['fɒs(ə)l; -sil] n. 化石
 adj. 化石的

anaerobic wastewater treatment 厌氧废水处理
biological agent 生物制剂
oxygen-free 无氧的
break down 分解
biodegradable material 生物降解材料
anaerobic digestion 厌氧消化
that is 即；就是说；换言之
methane gas 沼气

in addition to 除……之外
extracellular enzyme 细胞外酶
inhabiting bacteria 栖息细菌
fatty acid 脂肪酸
facultative and anaerobic bacteria 兼性厌氧菌
methanogenic bacteria 产甲烷菌
result in 导致；引起
take place 发生
organic acid 有机酸
carbon dioxide 二氧化碳
consist of 由……构成
result from 起因于；由……造成
power source 电源；能源
engine fuel 发动机燃料
in other words 换句话说
high concentration 高浓度
greenhouse gases 温室气体
contribute to 有助于
global warming 全球变暖
greenhouse effect 温室效应
lead to 导致
fossil fuel 矿物燃料；化石燃料

Notes

1. Anaerobic wastewater treatment uses biological agents in an oxygen-free environment to remove impurities from wastewater. 厌氧废水处理是利用生物制剂在无氧环境中去除废水中的污染物。

2. The biological agents used in the process are microorganisms that consume or break down biodegradable materials in sludge, or the solid portion of wastewater following its filtration from polluted water. 厌氧废水处理过程中使用的生物制剂为微生物，这些微生物能消耗或分解污泥中的生物降解材料以及污水过滤后的废水中的固体部分。

- the biological agents used in the process...used 为过去分词，此处用作后置定语修饰 the biological agents；过去分词可作后置定语修饰名词或名词短语，被修饰部分承受分词动作。

- ...wastewater following its filtration...following 为现在分词，此处作后置定语修饰

wastewater；现在分词可作后置定语修饰名词或名词短语，被修饰部分发出分词动作。

3. An excellent way to decrease the amount of organic matter leftover in things such as sewage and leftover food, anaerobic digestion is typically a component of any biological wastewater treatment system. 作为减少污水和剩余食物中有机残留物的一个极佳方法，厌氧消化是生物废水处理系统中具有代表性的一个组成部分。
   - such as 比如；诸如；像
     eg. Heroes such as Yang Jingyu will always live in the hearts of the people. 像杨靖宇这样的英雄人物，将永远活在人民的心里。

4. An anaerobic treatment system is a complex three-step process that produces methane gas (in addition to other products) from the biological digestion of sewage waste. 厌氧处理系统包含复杂的三步过程，过程中除了产生其他生成物之外，还会产生沼气。
   - ...that produces methane gas... 从 that 开始一直到句尾，为 that 引导的定语从句部分，修饰 a complex three-step process。

5. During the initial stages of the sludge breakdown, the microorganisms, which are mostly bacteria, convert the waste into organic acids, ammonia, hydrogen and carbon dioxide. In the final stages of anaerobic wastewater treatment, the remains of the sludge are converted, by a single-celled microorganism known as a methanogen, into a biogas consisting of methane and carbon dioxide. 在最初的污泥分解阶段，主要由细菌构成的微生物将废料转变为有机酸、氨气、氢气和二氧化碳。在厌氧废水处理的最后阶段，一种被称为产甲烷菌的单细胞微生物将污泥中的残余物转化成含有沼气和二氧化碳的生物气体。
   - covert...into... 把……转变为……
     eg. The signal will be converted into digital code. 信号将被转变成数字编码。
   - consist of 由……构成
     eg. The atmosphere consists of more than 70% of nitrogen. 大气中含有 70% 以上的氮气。

   Substances consist of small particles called molecules. 物质是由叫做分子的微粒组成的。

6. The biogas that results from the anaerobic wastewater treatment may actually be harnessed and used as an alternative power source for cooking, lighting, heating and engine fuel. 厌氧废水处理过程中产生的生物气体可作为替代能源用于做饭、照明、取暖和发动机燃料。
   - result from 由……引起；产生于……
     eg. Her blindness of both eyes resulted from a traffic accident. 她的双目失明是由于

一次交通事故造成的。

Earthquakes can result from stresses in the earth's crust. 地壳内的应力可能引起地震。

7. Many in the scientific community believe that high concentrations of methane and carbon dioxide, also known as greenhouse gases, in Earth's atmosphere, contribute to the process of global warming. 科学家们普遍认为地球大气层中高浓度的沼气和二氧化碳（又名温室气体）是导致全球变暖的元凶。
- contribute to 有助于；促成

    eg. Colloids contribute to the character of the soil in other ways. 胶质体对土壤的其他性状也有影响。

    The three sons also contribute to the family business. 这 3 个儿子也为家族事业作出了贡献。

Exercises

I. Answer the following questions after reading the text.

1. What are the functions of biological agents in anaerobic wastewater treatment?
2. Why is anaerobic wastewater treatment also called anaerobic digestion?
3. Please state three major steps in an anaerobic treatment system.
4. Where does the anaerobic process usually take place?
5. Please state additional benefits of anaerobic wastewater treatment.

II. Decide whether each of the following statements is true (T) or false (F) according to the text.

(　) 1. Anaerobic wastewater treatment uses biological agents in a non-oxygen environment to remove impurities from wastewater.

(　) 2. Anaerobic wastewater treatment is also known as anaerobic digestion due to the action of the organic matters.

(　) 3. Fatty acids result from macromolecules.

(　) 4. A single-celled microorganism known as a methanogen converts the remains of the sludge into a biogas consisting of methane, hydrogen and carbon dioxide.

(　) 5. An additional advantage of anaerobic wastewater treatment is its reduction of gas emissions.

III. Translate the following words or phrases into English.

1. 厌氧废水处理 _____
2. 生物制剂 _____
3. 生物降解材料 _____
4. 厌氧消化 _____
5. 兼性厌氧菌 _____
6. 脂肪酸 _____
7. 产甲烷菌 _____
8. 有机酸 _____
9. 二氧化碳 _____
10. 温室气体 _____

IV. Translate the following sentences into Chinese.

1. Anaerobic wastewater treatment is also known as anaerobic digestion due to the action of the microorganisms. That is, they are essentially "digesting" the polluted parts of the water.

2. The first stage is the hydrolysis of lipids, cellulose, and protein. Extracellular enzymes produced by the inhabiting bacteria break down these macromolecules into smaller and more digestible forms.

3. Next, these molecules are decomposed into fatty acids. This decomposition is performed by several facultative and anaerobic bacteria.

4. Finally, methanogenic bacteria digest these fatty acids, resulting in the formation of methane gas.

5. In other words, by capturing and utilizing the methane and carbon dioxide produced by anaerobic digestion, the biogas is not released into the atmosphere.

V. Match the words and expressions with their meanings.

(　) 1. extracellular enzyme　　　a. 栖息细菌
(　) 2. inhabiting bacteria　　　　b. 细胞外酶
(　) 3. power source　　　　　　c. 全球变暖
(　) 4. fossil fuel　　　　　　　　d. 矿物燃料

(　) 5. high concentration　　　　e. 发动机燃料
(　) 6. global warming　　　　　　f. 能源
(　) 7. greenhouse effect　　　　　g. 温室效应
(　) 8. engine fuel　　　　　　　　h. 高浓度

VI. Best choices.

1. Anaerobic wastewater treatment uses _____ agents to remove impurities.

A. biological　　　　　　B. physical

C. chemical　　　　　　　D. medical

2. Anaerobic wastewater treatment mainly uses microorganisms to digest the polluted parts of the water in an _____ environment.

A. oxygen　　　　　　　　B. oxygen-free

C. acid　　　　　　　　　D. alkaline

3. What is produced in the final stages of anaerobic wastewater treatment?

A. fatty acids　　　　　　B. protein

C. cellulose　　　　　　　D. methane gas

4. The biogas produced in the final stages consist of _____.

A. methane　　　　　　　B. carbon dioxide

C. hydrogen　　　　　　　D. both A and B

5. Many scientists think that _____ traps heat from the sun in the atmosphere, thereby increasing global temperature.

A. oxygen　　　　　　　　B. hydrogen

C. nitrogen　　　　　　　D. carbon dioxide

VII. Fill in the following blanks in the figure.

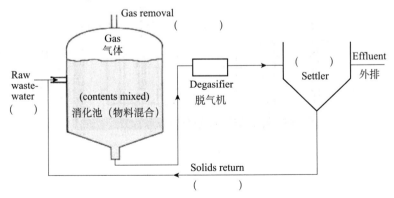

Text C Flocculation

Naturally occurring silt particles suspended in water are difficult to remove because they are very small, often colloidal in size, and possess negative charges, and are thus prevented from coming together to form large particles that could more readily be settled out. The removal of these particles by settling requires first that their charges be neutralized and second that the particles be encouraged to collide with each other. The charge neutralization is called coagulation, and the building of larger flocs from smaller particles is called flocculation.

Flocculation is one step in the water and wastewater treatment process. In a flocculation tank, the water is stirred or otherwise moved around so that the particles move around, bump into other particles, and stick to one another. Eventually the small and difficult to remove particles in the water form large clumps which can then be easily removed. Chemicals (most commonly "alum") are often added to the water going into a flocculation tank to help aid particle formation.

Alum (aluminum sulfate) is the usual source of trivalent cations in water treatment. Alum has an advantage in addition to its high positive charge: some fraction of the aluminum ions may form aluminum oxide and hydroxide by the reaction.

$$Al^{3+} + 3OH^- \rightarrow Al(OH)_3 \downarrow$$

These complexes are sticky and heavy and will greatly assist in the clarification of the water in the settling tank if the unstable colloidal particles can be made to come in contact with the floc. This process is enhanced through an operation known as flocculation.

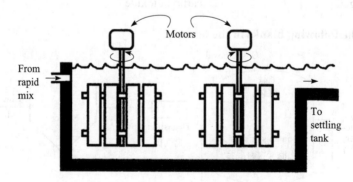

Figure 1 Flocculator used in water treatment

When the flocs have been formed they must be separated from the water. This is invariably done in gravity settling tanks that allow the heavier-than-water particles to settle to the bottom. Settling tanks are designed to approximate uniform flow and to minimize turbulence. Hence, the two critical elements of a settling tank are the entrance and exit configurations. Figure 2 shows one type of entrance and exit configuration used for distributing the flow entering and leaving the water treatment settling tank.

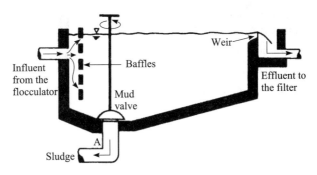

Figure 2　Settling tank used in water treatment

Words and Expressions

　　colloidal　[kə'lɒidəl] *adj.* 胶体的；胶质的；胶状的
　　settling　['setliŋ] *n.* 沉淀
　　neutralize　['nju:trəlaiz] *vi. vt.*（使）中和
　　collide　[kə'laid] *vi. vt.*（使）碰撞；（使）相撞
　　coagulation　[kəʊ,æɡjʊ'leiʃən] *n.* 凝聚
　　floc　[flɒk] *n.* 絮体
　　flocculation　[flɔkju'leiʃən] *n.* 絮凝
　　stir　[stɜ:] *vi. vt.* 搅动
　　bump　[bʌmp] *vi. vt.* 碰；撞
　　clump　[klʌmp] *n.* 丛；块
　　chemical　['kemik(ə)l] *n.* 化学制品；化学药品
　　alum　['æləm] *n.* 矾；明矾
　　aluminum　[ə'lu:minəm] *n.* 铝
　　sulfate　['sʌlfeit] *n.* 硫酸盐
　　trivalent　[trai'veil(ə)nt] *adj.* 三价的
　　cation　['kætaiən] *n.* 阳离子；正离子

fraction   ['frækʃ(ə)n] n. 部分
ion   ['aiən] n. 离子
oxide   ['ɒksaid] n. 氧化物
hydroxide   [hai'drɒksaid] n. 氢氧化物
complex   ['kɒmpleks] n. 复合物
clarification   [,klærifi'keiʃ(ə)n] n. 澄清；净化
gravity   ['græviti] n. 重力
minimize   ['minimaiz] vi. vt.（使）最小化
turbulence   ['tɜːbjʊl(ə)ns] n. 涡流
configuration   [kən,figə'reiʃ(ə)n; -gjʊ-] n. 结构；配置
distribute   [di'stribjuːt; 'distribjuːt] vt. 分配；散布
flow   [fləʊ] n. 流量

silt particle   粉砂颗粒
settle out   沉淀出来
negative charge   负电荷
charge neutralization   电荷中和
flocculation tank   絮凝池
stick to   粘住
particle formation   颗粒成型
aluminum sulfate   硫酸铝
positive charge   正电荷
aluminum oxide   氧化铝
assist in   帮助
gravity settling   重力沉降
entrance and exit   出入口
uniform flow   等速流

Notes

1. The charge neutralization is called coagulation, and the building of larger flocs from smaller particles is called flocculation. 这种电中和被称为凝聚，而由小颗粒形成大絮体的过程称为絮凝。
2. In a flocculation tank, the water is stirred or otherwise moved around so that the particles move around, bump into other particles, and stick to one another. 在絮凝池内，通过搅动或其他方式让水搅动起来，以便水中的颗粒能够相互碰撞，彼此吸附。

3. Chemicals (most commonly "alum") are often added to the water going into a flocculation tank to help aid particle formation. 在进入絮凝池的水里经常会加入一些化学品（最常见的是明矾），用以帮助形成大颗粒。
4. These complexes are sticky and heavy and will greatly assist in the clarification of the water in the settling tank if the unstable colloidal particles can be made to come in contact with the floc. 这些复合物有黏性，重量大，不稳定的胶体粒子如果能够和絮凝物相接触，则可以极大地促使沉淀池水的澄清。
5. Settling tanks are designed to approximate uniform flow and to minimize turbulence. Hence, the two critical elements of a settling tank are the entrance and exit configurations. 沉淀池应设计为近似等速流及能够将涡流最小化，因此，设计沉淀池的两个关键因素是出口和入口结构。

Exercises

I. Answer the following questions after reading the text.

1. Why is it difficult to remove silt particles suspended in water?
2. What are the requirements of removing silt particles in water?
3. Please state the process of flocculation.
4. What are the advantages of Alum in water treatment?
5. What are the design principles of a settling tank?

II. Decide whether each of the following statements is true (T) or false (F) according to the text.

(　) 1. Naturally occurring silt particles suspended in water possess positive charges, and are thus prevented from coming together to form large particles.
(　) 2. The charge neutralization is called flocculation.
(　) 3. The building of larger flocs from smaller particles is called coagulation.
(　) 4. Alum possesses positive charge.
(　) 5. Settling tanks are designed to approximate uniform flow and to minimize turbulence.

III. Translate the following words or phrases into English.

1. 胶体微粒　　　　　　　　　2. 负电荷

3. 正电荷　　　　　　　　　　4. 电荷中和

5. 絮凝池 _____   6. 硫酸铝 _____

7. 氧化铝 _____   8. 重力沉降 _____

9. 等速流 _____   10. 粉砂颗粒 _____

IV. Translate the following sentences into Chinese.

1. Naturally occurring silt particles suspended in water are difficult to remove because they are very small, often colloidal in size, and possess negative charges, and are thus prevented from coming together to form large particles that could more readily be settled out.

2. The removal of these particles by settling requires first that their charges be neutralized and second that the particles be encouraged to collide with each other.

3. Alum (aluminum sulfate) is the usual source of trivalent cations in water treatment.

4. Alum has an advantage in addition to its high positive charge: some fraction of the aluminum ions may form aluminum oxide and hydroxide by the reaction.

5. This is invariably done in gravity settling tanks that allow the heavier-than-water particles to settle to the bottom.

V. Best choices.

1. The building of larger flocs from smaller particles is called composition of pollutants in a(n) _____ environment.
   A. coagulation   B. flocculation   C. acid   D. alkaline

2. The most common chemical in coagulation is _____.
   A. aluminum sulfate   B. aluminum chloride
   C. hydrochloric acid   D. sodium hydroxide

3. When the flocs have been formed they must be separated from the water in a(n) _____.
   A. flocculator   B. aeration tank
   C. gravity settling tank   D. digesting tank

VI. Fill in the following blanks in the figures.

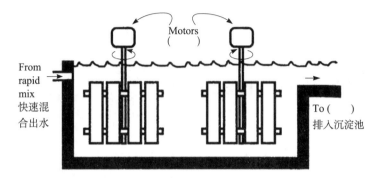

Figure 1　Flocculator used in water treatment

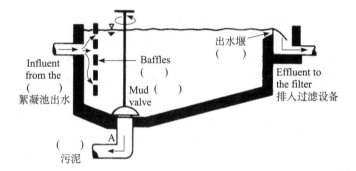

Figure 2　Settling tank used in water treatment

Unit 4 Tertiary Treatment

Text A Applications of Advanced Oxidation

Introduction

Many physical, biological, and chemical processes are used in wastewater treatment. But some contaminants found in wastewater are recalcitrant to some degree to commonly applied processes. Chemical oxidation processes are transformation processes that may augment current treatment schemes. Oxidation processes may destroy certain compounds and constituents through oxidation and reduction reactions.

Advanced oxidation is chemical oxidation with hydroxyl radicals（羟基自由基）, which are very reactive, and short-lived oxidants. The radicals need to be produced on site, in a reactor where the radicals can contact the organics in the wastewater. Hydroxyl radicals may be produced in systems using: ultraviolet radiation（紫外线照射）/hydrogen peroxide（过氧化氢）, ozone（臭氧）/hydrogen peroxide, ultraviolet radiation/ozone, Fenton's reagent（芬顿试剂）(ferrous iron and hydrogen peroxide), titanium dioxide（二氧化钛）/ultraviolet radiation, and through other means.

Application to WW Treatment

As shown in Figure 1, AOPs（advanced oxidation processes）may be used in wastewater treatment for (1) overall organic content reduction (COD), (2) specific pollutant destruction, (3) sludge treatment, (4) increasing bioavailability of recalcitrant organics, and (5) color and odor reduction.

Advanced oxidation was investigated for reducing the overall organic content, measured with chemical oxygen demand (COD), of a wastewater from an industrial facility that produced cleaners and floor care products. The wastewater contained up to 5% surfactants, solvents, and chelating agents（螯合剂）. The COD of the wastewater from the facility needed to be reduced in concentration before discharge to the local public treatment facility.

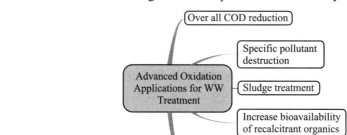

Figure 1 Application of AOPs for WW treatment

Bench-scale experiments were conducted to evaluate the potential for reducing the COD of this wastewater with Fenton's reagent. The results of the experiments showed that Fenton's reagent was very successful, reducing the COD more than 96% as shown in Figure 2. It was also found in these experiments that the reactions were exothermic (3.748 ± 0.332 J of heat were released per mg/L COD removed). This heat could be captured for useful purposes in the facility.

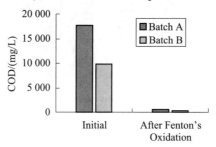

Figure 2 Reduction in COD of an industrial WW with Fenton's oxidation

Advanced oxidation can also be used to destroy specific pollutants that remain in wastewater after other treatment steps. Fenton's oxidation was used by Bergendahl et al. to decrease the concentration of organic contaminants（有机污染物）in water. Experiments with a pilot-scale system (shown in Figure 3) were undertaken and illustrated successful degradation of many of the contaminants present with Fenton's oxidation. Many of the contaminants present in this wastewater were significantly decreased in concentration (i.e. m- and p-xylenes), although some were not (i.e. 1,1,1- trichloroethane). Overall, there was an 81.8% reduction in organic contaminant mass in the water following Fenton's oxidation. Fenton's oxidation has also been found to be very effective for mineralizing methyl *tert-butyl* ether (MTBE) in water.

The destruction of endocrine disruptors, a class of contaminants that have recently been found in wastewater, with ozone was investigated by Nakagawa *et al.* using pilot-scale reactors. As shown in Figure 4, they were able to obtain significant reduction in estradiol（雌二醇）, bisphenol A（双酚 A）, and nonylphenol（壬基酚）concentrations with a 1 mg/L ozone dose, but no destruction in estrone（雌激素酮）. An ozone dose of 5 mg/L resulted in almost complete destruction of estradiol, bisphenol A, and nonylphenol, and only 20% reduction in estrone concentration. These experiments show that the effectiveness of advanced oxidation is dependent on the specific compound to be destroyed.

Figure 3　Pilot-scale Fenton's oxidation system used for destruction of organic contaminations in water

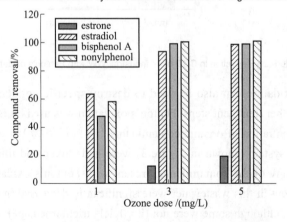

Figure 4　Destruction of endocrine disruptors with ozonation

The hydroxyl radicals produced by AOPs are also effective for treating and conditioning sludge produced from wastewater processes as they destroy cell walls of microorganisms. The cell material becomes solubilized with advanced oxidation, and amenable for further oxidation or other treatment. AOPs can be implemented into wastewater systems using two configurations as shown in Figure 5.

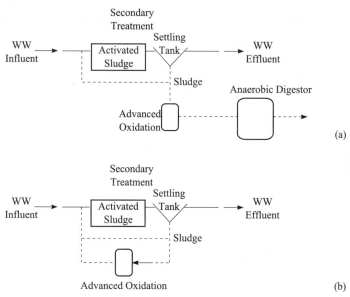

Figure 5 Two schemes for implementing AOPs in sludge treatment
(a) Advanced oxidation of sludge before anaerobic digestion
(b) Advanced oxidation of recycled sludge

Treatment scheme (b) in Figure 5 was investigated by Yasui and co-workers in a full-scale activated sludge system treating 120 000 gal/d municipal wastewater. Part of the recycled sludge in the system was subjected to ozonation. The WW treatment system was run for up to 10 months, with no excess sludge disposed — it was a sludgeless system.

New efforts have focused on integrating advanced oxidation with other technologies. Molecular sieve zeolites(分子筛沸石) have great capacity to adsorb organic contaminants from water. Yet these contaminants are merely transferred to the solid phase. However, advanced oxidation can destroy these adsorbed contaminants and regenerate the sorbent. Figure 6 shows preliminary experiments where silicalite, after repeated adsorption cycles with chloroform in water, becomes saturated and unable to sorb any more contaminant (after cycle 8). But after advanced oxidation, the silicalite regains its original sorption capacity (after cycle 9).

Conclusion

AOPs can be utilized in wastewater treatment for: overall organic content (COD) reduction, specific pollutant destruction, sludge treatment, increase of bioavailability of recalcitrant organics, and color and odor reduction.

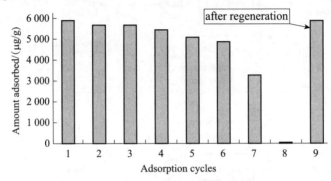

Figure 6   Regeneration of silicalite saturated with chloroform using Fe^o/H_2O_2 advanced oxidation. Regeneration only after 8th adsorption cycle

Words and Expressions

estradiol   [,estrə'daiəl] *n.* 雌二醇
estrone   ['estrəʊn] *n.* 雌激素酮
bisphenol   ['bisfinɒl] *n.* 双酚
bioavailability   [,baiəʊə,velə'biliti] *n.* 生物利用度
chelating   [ˈkiːleitiŋ] *v.* 螯化的；有螯的，螯合的（chelate 的现在分词）
hydroxyl   [hai'drɒksail; -sil] *n.* 羟基
ozone   ['əʊzəʊn] *n.* 臭氧
ultraviolet   [ˌʌltrə'vaiələt] *n.* 紫外线
chloroform   ['klɔːrəfɔːm; 'klɒr-] *n.* 氯仿
reagent   [ri'eidʒ(ə)nt] *n.* 试剂
silicalite   [sili'kælait] *n.* 硅质岩
titanium   [tai'teiniəm; ti-] *n.* [化学] 钛（金属元素）
xylenes   ['zaili:n] *n.* 二甲苯
p-xylenes   对二甲苯

m-xylenes 间二甲苯

hydroxyl radicals 羟基自由基
ultraviolet radiation 紫外线照射
hydrogen peroxide 过氧化氢
Fenton's reagent 芬顿试剂
titanium dioxide 二氧化钛
AOPs（advanced oxidation processes）高级氧化技术
chelating agents 螯合剂
organic contaminants 有机污染物
methyl tert-butyl ether (MTBE) 甲基叔丁醚
nonylphenol concentrations 壬基酚浓度
Molecular sieve zeolites 分子筛沸石

Notes

1. Many physical, biological, and chemical processes are used in wastewater treatment. But some contaminants found in wastewater are recalcitrant to some degree to commonly applied processes. 在废水处理过程中我们使用了很多物理、生物和化学的处理方法，但是普通的处理方法对于废水中的一些污染物在某种程度上是无效的。
   - to some degree 在某种程度上
   - found in wastewater 过去分词短语作后置定语，修饰名词 contaminants

2. Advanced oxidation was investigated for reducing the overall organic content, measured with chemical oxygen demand (COD), of a wastewater from an industrial facility that produced cleaners and floor care products. 我们研究高级氧化处理是为了减少废水中总的有机物含量，这种有机物含量可通过测量生产清洁剂和地板护理产品的工业设备所产生的废水的化学需氧量 (COD) 来测定。
   - Advanced oxidation was investigated for reducing the overall organic content. 为被动语态
   - measured with chemical oxygen demand (COD) 过去分词短语修饰 organic content
   - of a wastewater ... 为后置定语修饰 organic content
   - that produced cleaners and floor care products. 为定语从句，修饰前面的 industrial facility

3. Advanced oxidation can also be used to destroy specific pollutants that remain in wastewater after other treatment steps. 在其他处理完成后，高级氧化技术也可被用来破坏经过处理还残留在废水中的特定的污染物。

- that 引导定语从句修饰先行词 pollutants
- specific pollutants 特定污染物
4. These experiments show that the effectiveness of advanced oxidation is dependent on the specific compound to be destroyed. 这些实验表明高级氧化的有效性取决于要破坏的特定化合物。
   - that 引导宾语从句
   - to be destroyed 动词不定时作后置定语修饰前面的名词 compound
   - be dependent on 依靠；依赖
5. New efforts have focused on integrating advanced oxidation with other technologies. 我们把重点放在用其他技术整合高级氧化技术上。
   - focus on 集中于……
     eg. Our meeting focuses on the question of women's right. 我们会议的重点是妇女权利问题。

Exercises

I. Tell the differences between the two pictures.

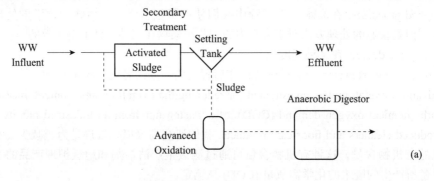

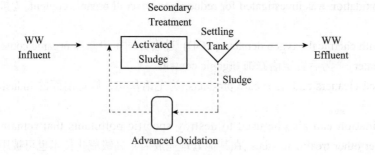

II. Best choices.

(   ) 1. Regeneration only after _____ adsorption cycle.
A. 6th   B. 7th   C. 8th   D. 9th

(   ) 2. After advanced oxidation, the silicalite regains its original _____ capacity.
A. surface   B. sorption   C. production   D. emission

(   ) 3. An ozone dose of _____ mg/L resulted in almost complete destruction of estradiol, bisphenol A, and nonylphenol.
A. 4   B. 2   C. 3   D. 5

(   ) 4. ...some contaminants found recalcitrant in wastewater are to some degree to commonly applied processes, "recalcitrant" means _____.
A. important   B. useful
C. dangerous   D. useless

(   ) 5. there was _____ reduction in organic contaminant mass in the water following Fenton's oxidation .
A. 81.8%   B. 61.8%   C. 71.8%   D. 51%

III. Decide whether each of the following statements is true (T) or false (F) according to the text.

(   ) 1. Oxidation processes may destroy certain compounds and constituents through oxidation and reduction reactions.

(   ) 2. Advanced oxidation can be used to produce specific pollutants that remain in wastewater after other treatment steps.

(   ) 3. The effectiveness of advanced oxidation is dependent on compounds.

(   ) 4. The cell material becomes solubilized with advanced oxidation, and amenable for further oxidation or other treatment.

(   ) 5. Molecular sieve zeolites can adsorb a few organic contaminants from water.

IV. Translate the following words or phrases into English.

1. 羟基自由基　　　　　　　2. 高级氧化技术
   _____   _____

3. 有机污染物　　　　　　　4. 废水处理
   _____   _____

5. 芬顿试剂　　　　　　　　6. 臭氧
   _____   _____

7. 浓度　　　　　　　　　　8. 紫外线照射
   _____   _____

9. 吸附循环   10. 三级水处理

V. Match the words and expressions with their meanings.

( ) 1. specific pollutant　　　　　　　　a. 氧化和还原反应
( ) 2. AOPS　　　　　　　　　　　　　b. 高级氧化技术
( ) 3. Molecular sieve zeolites　　　　　c. 二氧化钛
( ) 4. MTBE　　　　　　　　　　　　　d. 特定污染物
( ) 5. oxidation and reduction reactions　e. 雌激素酮
( ) 6. overall organic content　　　　　f. 甲基叔丁醚
( ) 7. Estradiol　　　　　　　　　　　g. 雌二醇
( ) 8. titanium dioxide　　　　　　　　h. 分子筛沸石
( ) 9. nonylphenol concentrations　　　i. 有机物总量
( ) 10. estrone　　　　　　　　　　　j. 壬基酚浓度

VI. Translate the following sentences into Chinese.

1. These experiments show that the effectiveness of advanced oxidation is dependent on the specific compound to be destroyed.

2. Advanced oxidation is chemical oxidation with hydroxyl radicals（羟基自由基）, which are very reactive, and short-lived oxidants.

3. The COD of the wastewater from the facility needed to be reduced in concentration before discharge to the local public treatment facility.

4. The hydroxyl radicals produced by AOPs are also effective for treating and conditioning sludge produced from wastewater processes as they destroy cell walls of microorganisms.

5. AOPs can be utilized in wastewater treatment for: overall organic content (COD) reduction, specific pollutant destruction, sludge treatment, increase of bioavailability of recalcitrant organics, and color and odor reduction.

Text B Advanced Waste Treatment of Secondary Effluent with Active Carbon

Introduction

The past decade has seen a fundamental change in our concepts of waste treatment and pollution control. Originally, the treatment of municipal wastes was primarily concerned with the preservation of public health. Next we become concerned with esthetic concepts such as the elimination of visible signs of pollution and the maintenance of oxygen levels for sustenance of marine life in receiving waters. In the last decade two new concepts have been developed. The first is that in many areas water is becoming scarce so that it may be necessary to use water more than once. The second is that natural waters should be maintained in a condition of purity so that the overall impact of water use on the environment is minimized. Both of these concepts require treatment of wastewaters that is fundamentally different from the treatment that was acceptable at a time when our only concerns were with public health, esthetics, and oxygen depletion. In this context activated carbon adsorption is the key unit process in the treatment of wastewater to produce effluents meeting our present requirement for effluent quality and receiving water preservation.

The use of coal in water treatment goes back to the last century. In 1883, 22 water plants(水处理厂) in the United States were reported to be employing charcoal filters. These were later abandoned because of the low adsorptive capacity （吸附能力） of the charcoal. The production of activated carbon （活性炭） was started in 1913 by a predecessor of Westvaco. However, its first recorded application to municipal water treatment was not until 1927 when two Chicago meat packing companies used powdered activated carbon(粉末活性炭) to remove tastes from their water supplies. During the 1930s, the use of powdered activated carbon to remove tastes and odors caused by traces of dissolved organics spread rapidly.

In 1960, the U. S. Public Health Service embarked on the Advanced Waste Treatment (废水深度处理) Research Program with two stated goals - to help abate water pollution problems and, more startling in concept, to renovate water for direct and deliberate reuse. The program focused early on adsorption as the most promising process for achieving its stated goals, and on activated carbon as the most feasible adsorbant（吸附剂）. A series of studies were commissioned by the Public Health Service and, since 1966, the Federal Water Pollution Control Administration, to evaluate the feasibility of activated carbon adsorption for wastewater renovation. These studies concentrated on two aspects - the physical configuration for the most economical use of the adsorptive properties（吸附性能）of the carbon and the reactivation（活化，再生）of the carbon for reuse.

Based on results of these studies, several demonstration plants（示范工厂）were designed to obtain data from commercial equipment, one of these plants is a joint effort of the County Sanitation Districts of Los Angeles County and the Federal Water Pollution Control Administration and is located at Pomona, California. The plant includes five carbon contactors and has a capacity of 400 gpm（每分钟加仑数）. A second plant is located at Lake Tahoe, California and is operated by the South Tahoe Public Utility District. The plant has a capacity of 7.5 mgd（百万加仑每天）. A third plant, located on Long Island, New York, is the subject of this paper.

Background

Nassau County occupies 291 sg. mi. of Long Island immediately adjacent to the City of New York. During the last two decades, the County has experienced an explosive growth of population and water consumption. Since the County's only source of water supply is the local ground water, whose safe yield is limited by its rate of recharge（补给率）, the continuation of this growth presages a crisis in water supply. Overpumping（过度抽取）results in lowering of the ground water levels and intrusion of salt water in the aquifer.

Development of the County has also decreased the rate of recharge of the ground water. The installation of public sewer system diverts wastewater previously recharged into the ground through septic tanks and cesspools to ocean outfalls. Present projections indicate that, if present trends continue, the net amount of water withdrawn from the aquifers will exceed the rate of recharge by 1977.

One plan to increase the permissible withdrawals is to create a hydraulic barrier（水力屏障）in the aquifer（含水层）. This barrier would prevent a natural outflow in the aquifer, estimated to be of the order of 30 mgd which is now lost to the sea. It would also prevent the intrusion（侵蚀）of salt water into the aquifer which is already becoming a problem in some areas of Nassau County. The barrier would be formed by injecting tertiary treated wastewater through a series of recharge wells along the southern perimeter of Nassau County.

Water Quality Requirements

In order to provide water of a quality necessary for injection into public water supply aquifers, the effluent of the existing sewage treatment plant must receive additional treatment to meet the following requirements:

1. U. S. Public Health Service Standards for drinking water.
2. Economical operation of injection system.
3. Chemical compatability with natural ground water.

The drinking water standard was adopted primarily in order to gain public acceptance of the concept of injecting treated wastewater into an aquifer which is used as a source of public water supply. Present plans provide for maintaining at least one mile separation between injection and water supply wells. This distance will insure that no particulates or bacteria will reach the water supply wells. Nevertheless, it was decided as a policy matter that the water as injected must meet the standards for drinking water.

Advanced Waste Treatment Process

The advanced waste treatment process used to achieve these water quality criteria consists of coagulation（混凝）with alum（明矾）, filtration（过滤）, adsorption（吸附）on activated carbon and disinfection with chlorine(加氯消毒).

Standards of water quality for economical operation of the injection system are being developed as part of the demonstration project. From injection tests conducted thus far, it is apparent that particulates must be maintained at the lowest possible level. Turbidity levels（浊度）of less than 0.5 J.T.U. appear to be desirable. Low levels of dissolved gases（溶解性气体）were considered desirable during the early stages of the project but do not appear to be as critical as they were thought to be. Turbidities in excess of 1.0 Jackson Units result in rapid buildup of pressure required to inject at a given rate of flow.

The principal problems of compatibility involve iron and phosphate concentrations. Iron precipitates in the aquifer, causing irreversible clogging of the formation. The role of phosphates is not yet fully understood. However, changes in phosphate concentration between water injected and injected water recovered have been observed, leading to the conclusion that phosphates interact with the fine clayey sands that comprise the aquifer.

Effluent from the final sedimentation tanks of the Bay Park Sewage Treatment Plant is pumped into a clarifier, where alum and coagulant（絮凝剂）aids are added. Sludge（污泥）recirculation is employed to improve coagulation and overcome sudden changes in water quality. Flow then passes by gravity to two mixed media filters operated in parallel, each containing a 36-inch bed of anthracite above a 12-inch layer of sand. Filter backwash is automatic, and includes facilities for air scour, surface wash, and high and low rate backwashing.

Filter effluent is pumped through four granular activated carbon adsorbers operating in series. Adsorber piping is arranged so that the order of the vessels can be rotated to change the sequence of flow and insure the most efficient utilization of carbon. Upon exhaustion, carbon is moved hydraulically to a regeneration system. Here the carbon is restored to its original activity by controlled burning off of the adsorbed organics in a multi-hearth furnace.

The renovated water is disinfected with chlorine prior to being pumped about one half mile to the test injection site. The injection facilities consist of a storage tank, degasifier（脱气塔）for removal of residual chlorine and dissolved gases, injection and redevelopment pumps, the injection well and 12 observation wells. The injection well is 36 inches in diameter by 500 ft. deep, and contains an 18 inch casing which supports a 16 inch screen set between elevations −420 and 480 ft. The annular space surrounding the screen has been backfilled with graded sand and contains an observation-well and geophysical probes. Other observation wells are located up to 200 ft. from the injection well.

Words and Expressions

adsorption　[æd'sɔ:pʃən]　*n.* 吸附
adsorbent　[əd'zɔ:bənt]　*n.* 吸附剂
overpumping　[əʊvə'pʌmpiŋ]　*n.* 过度抽取
preservation　[prezə'veiʃ(ə)n]　*n.* 保存；保留
reactivation　[ri(:)ækti'veiʃən]　*n.* 活化；再生
aquifer　['ækwifə]　*n.* 蓄水层
effluent　['efluənt]　*n.* 出水，污水
intrusion　[in'tru:ʒ(ə)n]　*n.* 侵蚀
coagulation　[kəʊ,ægjʊ'leiʃən]　*n.* 混凝
alum　['æləm]　*n.* 明矾
filtration　[fil'treiʃn]　*n.* 过滤
coagulant　[kəʊ,ægjʊ'leiʃən]　*n.* 絮凝剂
degasifier　[di'gæsifaiə]　*n.* 脱气塔
phosphate　['fɒsfeit]　*n.* 磷酸盐
turbidity　[tɜ:'bidəti]　*n.* 浊度；浑浊度
gpm　*abbr.* 每分钟加仑数（gallons per minute）
mgd　*abbr.* 百万加仑每天（million gallons per day，测量水流量的单位）

adsorptive capacity 吸附能力
activated carbon 活性炭
advanced waste treatment 废水深度处理
adsorptive properties 吸附性能
demonstration plants 示范工厂
dissolved gases 溶解性气体
rate of recharge 补给率
disinfection with chlorine 加氯消毒
hydraulic barrier 水力屏障
Jackson Units 杰克逊浊度单位
powdered activated carbon 粉末活性炭
redevelopment pumps 再生泵
screen set 网筛
secondary effluent 二级出水
turbidity levels 浊度
water plant 水处理厂

Notes

1. Originally, the treatment of municipal wastes was primarily concerned with the preservation of public health. Next we become concerned with esthetic concepts such as the elimination of visible signs of pollution and the maintenance of oxygen levels for sustenance of marine life in receiving waters. 最初，城市废物的处理首要关注维护公共健康。然后我们更加关注城市的美观，例如消除可见的污染痕迹，维持水域海洋生物生存的氧气水平。
   - marine life 海洋生物
   - waters 水域
   - concerned with 关心；涉及；与……有关
     eg. They are concerned with the training curriculum, but not the details of each course. 他们关注培训课程，但并不干预每一门课程的细节。

2. In this context activated carbon adsorption is the key unit process in the treatment of wastewater to produce effluents meeting our present requirement for effluent quality and receiving water preservation. 由此而论活性炭吸附是废水处理的关键过程，它可以进行二级出水深度处理以满足当前对废水质量和水保持的要求。
   - effluent 出水；废水。这里指二级出水后的深度处理
   - receiving water preservation 水保持

- meeting　作后置定语修饰前面的名词 effluents

3. The program focused early on adsorption as the most promising process for achieving its stated goals, and on activated carbon as the most feasible adsorbant. 作为达到既定目标的最有希望的方法，这个项目早期重点在于研究吸附，研究最可行的吸附剂，这个项目的重点在于研究活性炭。
   - as　代词，代替 the program
   - stated goal　既定目标
   - focus on　集中于

     eg. Our meeting focuses on the question of women's right. 我们会议的重点是妇女权利问题。

4. Present projections indicate that, if present trends continue, the net amount of water withdrawn from the aquifers will exceed the rate of recharge by 1977. 目前的推测表明，如果按照现在的趋势发展下去，从蓄水层取出的水的净含量到 1977 年将会超过地下水的再生量。
   - that　引导宾语从句
   - if　引导条件状语从句
   - withdrawn　过去分词短语用作后置定语，修饰名词 water
   - indicate *vt.*　表明；指出

     eg. Then we want to indicate that meter to the performer. 然后我们要把那个拍子向演奏者说明。

     The report today indicates a change in United States policy. 今天的报道表明了美国政策的变化。

5. The barrier would be formed by injecting tertiary treated wastewater through a series of recharge wells along the southern perimeter of Nassau County. 在纳苏县南部周边一系列的回灌井中注入三级处理废水就可以形成这道屏障。
   - tertiary treated wastewater　三级处理废水
   - recharge well　回灌井
   - along　引导介词短语后置作定语，意为"沿着……"
   - a series of　一系列

     eg. We learn language through a series of prompts and feedback. 我们通过一系列的提示和反馈学习语言。

6. Nevertheless, it was decided as a policy matter that the water as injected must meet the standards for drinking water. 尽管如此，作为政策的关键就是注入水必须要符合饮用水的标准。
   - nevertheless　然而；不过；虽然如此

eg. Nevertheless, the difference should not be ignored. 然而，这样的差距不应被忽略。
- as a policy 中 as 作介词用，意为"作为"
- it 作形式主语，真正的主语为 that 引导的从句

7. Iron precipitates in the aquifer, causing irreversible clogging of the formation. 铁在蓄水层中沉淀，会形成不可消除堵塞。
   - irreversible clogging   不可消除的堵塞
   - causing   引导现在分词短语作结果状语

Exercises

I. Best choices.

1. The production of activated carbon was started in _____ by a predecessor of Westvaco.
   A. 1913   B. 1914   C. 1923   D. 1935

2. Activated carbon adsorption is the _____ unit process in the treatment of wastewater.
   A. first   B. normal   C. key   D. last

3. Present plans provide for maintaining at least _____ separation between injection and water supply wells.
   A. two kilometers B. two miles   C. one kilometers D. one mile

4. Filter effluent is pumped through _____ granular activated carbon adsorbers operating in series.
   A. four   B. three   C. two   D. one

5. It was decided as a policy matter that the water as injected must meet the standards for _____ water.
   A. urban   B. drinking   C. polluted   D. waste

II. Decide whether each of the following statements is true (T) or false (F) according to the text.

(   ) 1. In 1960, the U. S. Public Health Service embarked on the Advanced Waste Treatment Research Program with two stated goals.

(   ) 2. The renovated water is disinfected with chlorine worse than being pumped about one half mile to the test injection site.

(   ) 3. Here the carbon is restored to its original activity by controlled burning off of the adsorbed organics in a multi-hearth furnace.

(　　) 4. The first recorded application of activated carbon to municipal water treatment was in 1927.

(　　) 5. The principal problems of compatibility involve iron and phosphate concentrations.

III. Traslate the following words or phrases into English.

1. 蓄水层 _____

2. 过度抽取 _____

3. 饮用水 _____

4. 活性炭 _____

5. 出水 _____

6. 二级出水 _____

7. 海洋生物 _____

8. 吸附剂 _____

9. 公共健康 _____

10. 水处理厂 _____

IV. Match the words and expressions with their meanings.

(　　) 1. recharge well　　　　　　　　a. 活性炭粉末

(　　) 2. powdered activated carbon　　b. 水力屏障

(　　) 3. Jackson Units　　　　　　　　c. 水保持

(　　) 4. receiving water preservation　　d. 浊度

(　　) 5. gpm　　　　　　　　　　　　e. 加氯消毒

(　　) 6. mgd　　　　　　　　　　　　f. 杰克逊浊度单位

(　　) 7. adsorptive properties　　　　　g. 百万加仑/天

(　　) 8. turbidity levels　　　　　　　　h. 回灌井

(　　) 9. hydraulic barrier　　　　　　　i. 加仑/分钟

(　　) 10. disinfection with chlorine　　　j. 吸附性能

V. Translate the following sentences into Chinese.

1. The advanced waste treatment process used to achieve these water quality criteria consists of coagulation with alum, filtration, adsorption on activated carbon and disinfection with chlorine.

2. Originally, the treatment of municipal wastes was primarily concerned with the preservation of public health.

3. The continuation of this growth presages a crisis in water supply.

4. From injection tests conducted thus far, it is apparent that particulates must be maintained at the lowest possible level.

5. Turbidities in excess of 1.0 Jackson Units result in rapid buildup of pressure required to be injected at a given rate of flow.

VI. Supplementary reading comprehension.

Carbon Adsorption System（活性炭吸附系统）

The design of a carbon adsorption system for the treatment of wastewaters involves consideration of the following parameters:

Type of carbon—granular or powdered.

Physical configuration（物理结构）—upflow or downflow, or mixed number ofstages, parallel or series, packed bed or expanded bed, external regeneration or continuous flow.

Carbon capacity—detention time, dosage rate.

Method of operation—pure adsorption, filtration, biochemical.

For the Nassau County project, granular carbon was selected over powdered carbon primarily because of the state of the art of carbon regeneration. Powdered carbon has some advantages over granular carbon.

Its initial cost is lower, 7.5 cents per pound against 30 cents for granular carbon. It reacts faster and more completely, and its dosage can be adjusted to meet changes in the composition of the influent to the system. On the other hand, even the cost of powdered carbon is not sufficiently low to permit its discard after a single use. Some experimental work is now in progress on powdered carbon regeneration, but it has not yet reached the stage where a full scale demonstration plant can be designed. Dewatering and incineration are the most feasible methods of disposal of waste powdered carbon.

Granular carbon has been in industrial use for many years and the technology for its regeneration is well established. It has the additional advantage of providing a margin of safety in operation that powdered carbon does not provide. Sudden changes in influent composition are common in wastewater treatment. If the dosage of powdered carbon is not adjusted to meet these changes, the effluent quality will reflect the insufficient dosage. Granular carbon has the capacity to withstand substantial changes in the influent composition with a much reduced effect on the effluent quality. This aspect and the availability of the regeneration technology were the major factors in the selection of granular carbon for the Bay Park project.

Even after the choice has been made between granular and powdered – carbon, some further selectivity is required. Activated carbons are manufactured from a variety of raw materials such as coal, wood, nut shells and pulping wastes. A carbon that must undergo multiple regenerations must have the capability of being handled with a minimum of deterioration or abrasion. Since coal derived carbons are harder and denser than other carbons, this type of carbon was specified for the Bay Park project.

As a result of operating experience, the additional requirement that the carbon contain less than 0.5% of iron by weight has been added. The limit on the iron also forced a change in the gradation, so that the specifications for carbon could be met with a commercially available product. The original carbon had a size range of 8 × 30 (passing a standard No. 8 mesh sieve, but retained on a No. 30 sieve). The replacement carbon has a size range of 14 ×40.

A number of physical configurations have been suggested for activated carbon adsorption systems. These in clude upflow-expanded bed（上向流扩张床）, upflow-compacted bed, downflow-single stage, downflow-multistage, and a quasi-countercurrent system（准逆流系统）, in which the flow is down in the first unit and up in the second unit, exhausted carbon being continuously removed in the first unit and regenerated or makeup carbon being added continuously in the second.

Upflow systems have the advantages of being less susceptible to plugging and more adaptible to continuous countercurrent operations, which in theory yield the most efficient carbon utilization. Downflow systems（下向流系统）require periodic backwashing to prevent the buildup of headloss and multistaging to approach countercurrent operation. The differences in equipment costs are of a second order compared with the costs of regeneration and makeup. Downflow systems are mechanically simpler and have greater flexibility as to rates of flow that can be applied. For the Nassau County project, a four-stage downflow system was selected. The four vessels containing the carbon are piped so that they are in series, with each unit capable of being the lead unit. In normal operation, the flow is applied to the vessel containing carbon closest to exhaustion. As it passes from unit to unit, it encounters successively more active carbon, until in the last unit it passes through the most recently regenerated carbon.

When the organic content of the product water starts to exceed the desired level, the first unit is taken off the line（离开工作线）and the carbon in this unit is transferred hydraulically to the dewatering tank of the carbon regeneration system. As soon as the transfer is completed, regenerated and makeup carbon from the storage tank is pumped back into the unit. The unit is then put back on the line, but in the last position in the sequence. In this manner, the countercurrent mode of the operation is maintained.

Laboratory bench and pilot plant studies were relied upon to furnish other design data. Laboratory bench studies were used to derive adsorption isotherms（吸附等温线）, which give some indications as to carbon dosage（炭的吸附容量）. Column tests were then used to determine the required contact time. The hydraulic loading then becomes a matter of convenience for the design of the equipment. For the Nassau County project, the following design parameters were adopted, based on over a year's pilot plant operations:

Total contact time (empty bed volume) 24 min.

Hydraulic Loading (approach velocity) 7.5 gpm/sq ft.

The combination of these factors resulted in a vessel diameter of 8 ft and a bed depth of 6 ft in each vessel. Each of the vessels contains 300 cuft or about 9 000 lb of carbon. The rate of exhaustion has been about 800 gal per pound of carbon or 1.25 lb per 1000 gal treated.

Economics

Unit costs for the advanced waste treatment process are given in the following table. The table is based on a COD reduction 90% from 50 mg/L in the secondary effluent to 5 mg/L in the product water and a phosphate reduction of 90% from 30 mg/L (as PO_4^{3-}) to 3 mg/L.

Estimated Unit Costs

Cents per 1000 qal	Plant capacity		
	1 / (mg/d)	10 / (mg/d)	100 / (mg/d)
Process costs, less labor coagulation	4.9	3.5	3.2
Filtration	1.8	1.1	1.0
Carbon adsorption	6.3 / 13.0	4.5 / 9.1	4.0 / 8.2
Operating labor	28.0 / 41.0	5.6 / 14.7	1.8 / 10.0

Annual charges have been assumed at 8.5% of the capital costs and include both debt service and an allowance for maintenance, repair and replacement. Unit costs also assume continuous operation at design capacity (100% load factor). The costs are for treatment only and do not include transmission or injection facilities(传输和注入设备).

Conclusions

The Nassau County project is demonstrating the feasibility of treating secondary effluent with a physiochemical process (物理化学处理) sequence involving activated carbon to remove organics resistant to biological treatment. The product water meets U.S. Public Health Service Standards for drinking water and has physical properties such as turbidity, color or odor equivalent to those of the domestic water supply. It can be recharged into the ground without causing any deterioration of the aquifer. Based on test operations now in progress, it is believed that the concept of using treated wastewater for hydraulic barriers against seawater intrusion is technically feasible.

The project opens up new potentials for water reuse in areas where fresh water supplies are scarce. Wastewater is always available where there are public water supply and sewerage systems. Newly adopted water quality standards will require many communities to provide more than conventional secondary treatment. With activated carbon treatment, the product water can now be made available for many forms of beneficial reuse requiring high quality water.

Decide whether each of the following statements is true (T) or false (F) according to the material.

( ) 1. Initial cost of powdered carbon is higher, 30 cents per pound against 7.5 cents for granular carbon.

( ) 2. Dewatering and incineration are the most feasible methods of disposal of waste powdered carbon.

( ) 3. Powdered carbon has been in industrial use for many years and the technology for its regeneration is well established.

( ) 4. As a result of operating experience the additional requirement that the carbon contain less than 1.5% of iron by weight has been added.

( ) 5. Newly adopted water quality standards will require many communities to provide more than conventional secondary treatment.

Unit 5　Sewage Treatment Plant

Text A　Typical Project of Sewage Treatment Plant (1)

　　Sewage treatment is the process of removing contaminants from wastewater and household sewage, both runoff, domestic, and commercial. It includes physical, chemical, and biological processes to remove physical, chemical and biological contaminants. Its objective is to produce an environmentally safe fluid waste stream (or treated effluent) and a solid waste (or treated sludge) suitable for disposal or reuse.

　　Sewage can be treated close to where it is created, or be collected and transported by a network of pipes and pump stations to a municipal treatment plant.

　　Sewage treatment generally involves three stages, called primary, secondary and tertiary treatment.

　　Primary treatment consists of temporarily holding the sewage in a quiescent（不动的；静止的）basin where heavy solids can settle to the bottom while oil, grease and lighter solids float to the surface. The settled and floating materials are removed and the remaining liquid may be discharged to secondary treatment.

　　Secondary treatment removes dissolved and suspended biological matter. Secondary treatment is typically performed by water-borne micro-organisms in a managed habitat. Secondary treatment may require a separation process to remove the micro-organisms from the treated water prior to discharge or tertiary treatment.

　　Tertiary treatment is sometimes defined as anything more than primary and secondary treatment in order to allow rejection into a highly sensitive or fragile ecosystem. Treated water is sometimes disinfected chemically or physically prior to discharge into a stream, river, bay（海湾）, lagoon（潟湖）or wetland（湿地，沼泽）. If it is sufficiently clean, it can also be used for groundwater recharge or agricultural purposes.

　　The sewage treatment plant in Changsha Environmental Protection Vocational College is a typical project. The treatment plant was designed and constructed for students' skill training by Environmental Engineering Department of this college in 2002. Three widely used processes, which include bio-contact oxidation, dissolved air flotation and flocculation technology, have been adopted in this three stages wastewater treatment system.

The effluent running out from the third teaching building and the office building in the college can be collected together as raw wastewater and transported to this sewage treatment plant.

Raw water quality as follows:

Table 1 Raw Water Quality (mg/L)

COD_{Cr}	BOD_5	SS	TN	TP
400	200	220	40	8

Equipment capacity as designed: wastewater quantity 8 m^3/h.

The designing effluent quality:

Table 2 Effluent Quality (mg/L)

Item	COD_{Cr}	BOD_5	SS	Petroleum	Total number of bacteria (in 1 liter)
Maximum	16.8	2.8	17.2	0.3	<10
Minimum	8.4	1.4	7.6	0.2	<10
Average value	12.5	2.2	10.3	0.25	<10
National level of discharging standard	100	30	70	10	—
Miscellaneous Urban Water Quality Standard GB/T 18920—2002	50	10	10	—	100
Treatment efficiency (%)	96.4	98.5	98.4	96.6	99.9

Process flows: Multistage (分级) and parallel (并联) connection have been adopted in this wastewater treatment system. Switch valves and drain-outlet pipes are reserved for the maneuverability of process switching.

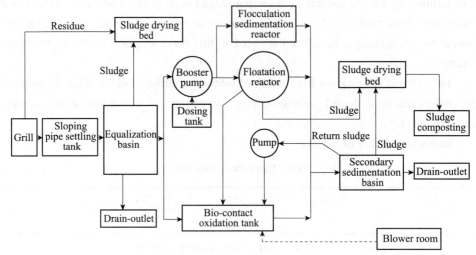

Figure1　Process flows of the sewage treatment plant in Changsha Environmental Protection Vocational College

Screening (Grill trough):

The influent sewage water passes through a bar screen to remove all large objects like cans（易拉罐；金属罐）, rags（破布）, sticks（树枝，棍子）, plastic packets etc. carried in the sewage stream. This is most commonly done with an automated mechanically raked bar screen in modern plants serving large populations, while in smaller or less modern plants, like the wastewater treatment system in our campus, a manually cleaned screen may be used. The raking action of a mechanical bar screen is typically paced according to the accumulation on the bar screens and flow rate. The solids are collected and later disposed in a landfill, or incinerated. Bar screens or mesh screens of varying sizes may be used to optimize solids removal. If gross solids are not removed, they become entrained（夹带）in pipes and moving parts of the treatment plant, and can cause substantial damage and inefficiency in the process.

Grit and grease removal (Sloping pipe settling tank):

In sloping pipe settling tank of our campus wastewater treatment system, the velocity of the incoming sewage is adjusted to allow the settlement of sand, grit, stones, and broken glass. These particles are removed because they may damage pumps and other equipment. For small sanitary sewer systems, the grit chambers may not be necessary, but grit removal is desirable at larger plants.

In the wastewater treatment system in our campus, fat and grease are removed by passing the sewage through the sloping pipe settling tank where skimmers collect the fat floating on the surface. In modern plants, air blowers in the base of the tank may also be used to help recover the fat as froth. Many plants, however, use primary clarifiers with mechanical surface skimmers for fat and grease removal.

Grit chambers come in 3 types: horizontal grit chambers, vertical grit chambers and radial-flow grit chambers.

Flow equalization:

Equalization basins are used for temporary storage of diurnal （白天的） or wetweather flow peaks. Basins provide a place to temporarily hold incoming sewage during plant maintenance and a means of diluting and distributing batch （批次） discharges of toxic or high-strength waste which might otherwise inhibit biological secondary treatment.

Flow equalization basins in some other treatment plants require variable discharge control, typically include provisions for bypass and cleaning, and may also include aerators. Cleaning may be easier if the basin is downstream（下游的）of screening and grit removal.

Words and Expressions

pump　[pʌmp]　*n.* 泵；抽水机
　　　　　　　　vt. 打气；用抽水机抽……
　　　　　　　　vi. 抽水
construct　[kən'strʌkt]　*vt.* 建造；修建；建设；创立
disinfect　[disin'fekt]　*vt.* 将消毒；杀菌；灭菌
drain　[drein]　*n.* 排水；排水管；下水道；消耗
　　　　　　　vt. 排掉；使流出；喝光；耗尽
　　　　　　　vi. 排水；流干
maneuverability　[mə,nʊvə'biləti]　*n.* 可控性；可操作性
rake　[reik]　*n.* 耙子；斜度
　　　　　　vt. vi. 用耙子耙；倾斜
landfill　['lænd'fil]　*n.* 垃圾填埋场；垃圾堆
incinerate　[in'sinəreit]　*vt. vi.* 焚化
optimize　['ɒptimaiz]　*vt. vi.*（使）最优化；（使）完善
inefficiency　[,ini'fiʃənsi]　*n.* 低效；无效
skimmer　['skimə]　*n.* 撇油器；撇渣器
froth　[frɒθ]　*n.* 泡沫
bypass　['baipɑːs]　*n.* 旁流；旁路

inhibit　[in'hibit]　*vt.* 抑制；约束；禁止
aerator　['eiəreitə]　*n.* 曝气器

water-borne micro-organism　水生微生物
typical project　样板工程；典型项目
sewage treatment plant　水处理平台；污水处理厂
groundwater recharge　地下水补给；地下水补注
raw water quality　原水水质
equipment capacity　设备能力；设备容量
effluent quality　出水水质
process flow　操作流程
sloping pipe settling tank　斜管沉淀池
equalization basin　均化池
sludge drying bed　污泥干化池；晒渣池
drain-outlet　排水出口；排水孔
booster pump　加压泵
dosing tank　污水量配池
bio-contact oxidation tank　生物接触氧化池
floatation reactor　气浮反应器
flocculation sedimentation reactor　絮凝沉淀反应器
return sludge　回流淤泥
secondary sedimentation tank　二次沉淀池
sludge composting　淤泥堆肥
blower room　鼓风机室
grill trough　格栅槽
bar screen　隔栅
mesh screen　网筛
substantial damage　实质损害
grit and grease removal　沙砾与油脂去除
horizontal grit chamber　平流沉砂池
vertical grit chamber　竖流沉砂池
radial-flow grit chamber　径向流沉砂池
flow equalization　均流
temporary storage　暂时储存

Notes

1. Its objective is to produce an environmentally safe fluid waste stream (or treated effluent) and a solid waste (or treated sludge) suitable for disposal or reuse. 其目的在于通过处理后产生对环境安全的适用于进一步处置和再利用的废水（处理过的出水）和固体废弃物（处理过的淤泥）。
   - objective　目的；目标
   - environmentally safe　对环境安全的；环保的
   - fluid　液体；流体；流动的
   - effluent　流出物；污水；废气；流出的；发出的
     eg. effluent treatment, effluent disposal　废水处理
     　　industrial effluent　工业废液；工业流出物
     　　plant effluent　工厂废水
     　　effluent standard　排放标准；排污标准

2. Sewage can be treated close to where it is created, or be collected and transported by a network of pipes and pump stations to a municipal treatment plant. 人们可以在产生污水的地方就近进行污水处理，或者把污水通过管道和泵站网络收集并运送到城市污水净化厂进行处理。
   - pump station　泵站

3. Tertiary treatment is sometimes defined as anything more than primary and secondary treatment in order to allow rejection into a highly sensitive or fragile ecosystem. 有时候我们把三级处理定义为在初级处理和二级处理后继续截留废水进行深度处理后，使之进入极度敏感脆弱的生态系统的所有的废水处理过程。
   - be defined as　被定义为……；被称为……
     eg. Self-esteem can be defined as a confidence and satisfaction in oneself. 自尊可以被定义为对自己的自信和满意。
   - allow rejection into　不让进入……

4. Three widely used processes, which include bio-contact oxidation, dissolved air flotation and flocculation technology, have been adopted in this three stages wastewater treatment system. 在此三级递进的废水处理系统中采用了生物接触氧化、溶气浮选和絮凝这三种广泛应用的废水处理技术。
   - bio-contact oxidation　生物接触氧化
   - dissolved air flotation　溶气浮选
   - flocculation　絮凝
   - adopt　采用；采取

eg. We should adopt an effective economy measure. 我们应当采取一项有效的节约措施。

5. Switch valves and drain-outlet pipes are reserved for the maneuverability of process switching. 转换阀和出水管是被留作随机掌控操作流程转换用的。
   - switch valve　转换阀；切换阀；转换开关
   - drain-outlet pipe　出水管
   - reserve　保留；储备
   - maneuverability　可操作性；可控性；机动性

6. This is most commonly done with an automated mechanically raked bar screen in modern plants serving large populations, whilst in smaller or less modern plants, like the wastewater treatment system in our campus, a manually cleaned screen may be used. 在为大量人口服务的现代化的污水处理厂，人们最常应用的是自动化的斜耙格栅，而像我校的这种规模较小的不那么现代化的污水处理系统使用手动的净化格栅即可。
   - 现在分词短语 serving large populations 作后置定语修饰 plants
   - whilst　相当于 while，并列连词，连接前后两个句子
   - automated　自动化的；机械化的
   - mechanically　机械地
   - manually　手动的
   - This is done with　这是用……完成的
     eg. This is done with two tests. 这是通过两个测试完成的。

7. Bar screens or mesh screens of varying sizes may be used to optimize solids removal. 人们用不同尺寸的格栅或筛网来优化固体物质的去除。

8. For small sanitary sewer systems, the grit chambers may not be necessary, but grit removal is desirable at larger plants. 对于小的污水处理系统，并没必要采用沉砂池，不过大一些的污水处理厂就需采用沉砂去除了。
   - grit chamber　沉砂池

9. Basins provide a place to temporarily hold incoming sewage during plant maintenance and a means of diluting and distributing batch（批次）discharges of toxic or high-strength waste which might otherwise inhibit biological secondary treatment. 均化池既可在水处理平台维修时临时保存注入的污水，也是一种过滤和分配有毒的或含有高浓度废弃物的废水排放批次的工具，否则这些过量的有毒物质或高浓度废弃物会影响二级生物处理的效果。
   - basin　这里指的是均化池（equalization basin），是用以尽量减小污水处理厂进水水量和水质波动的构筑物，又称调节池。均化池按照所起的作用可以分为水量均化池（简称均量池）：主要起均化水量作用；水质均化池（简称均质池）：主要

起均化水质作用
- temporarily 临时地
- temporary 临时的；暂时的
- incoming 引入的；进入的
- maintenance 维护；维修；保养
- high-strength 高浓度的

10. Flow equalization basins in some other treatment plants require variable discharge control, typically include provisions for bypass and cleaning, and may also include aerators. 在其他的污水处理厂中，均流池还需要配备能进行可变流量控制的装置，典型的包括分流装置、清洗装置，也许还包括曝气机。
    - flow equalization basin 均流池
    - aerator 曝气机；充气机；通风装置
    - provisions 提供；供给

Exercises

I. Best choices.

1. Sewage treatment includes physical, chemical, and _____ processes to remove physical, chemical and biological contaminants.
    A. natural   B. agricultural   C. industrial   D. biological

2. Sewage treatment generally involves three stages, called primary, secondary and _____ treatment.
    A. firstly   B. tertiary   C. finally   D. fourthly

3. The sewage treatment plant in Changsha Environmental Protection Vocational College can work _____ .
    A. temporarily   B. rarely   C. efficiently   D. inefficiency

4. The sewage treatment plant in Changsha Environmental Protection Vocational College was designed and _____ in 2002.
    A. built   B. bought   C. brought   D. provide

5. Cleaning may be _____ if the basin is downstream of screening and grit removal.
    A. harder   B. faster   C. less difficult   D. quicker

II. Translate the following words or phrases into English.

1. 水处理平台；污水处理厂 _____
2. 排水出口；排水孔 _____
3. 撇油器；撇渣器 _____
4. 排放标准 _____
5. 回流淤泥 _____
6. 曝气器 _____
7. 隔栅 _____
8. 格栅槽 _____
9. 杀菌 _____
10. 去脂 _____

III. Match the words and expressions with their meanings.

( ) 1. typical project   a. 平均值
( ) 2. floatation reactor   b. 处理功效
( ) 3. equalization basin   c. 鼓风机室
( ) 4. average value   d. 污水量配池
( ) 5. blower room   e. 平流沉砂池
( ) 6. dosing tank   f. 絮凝沉淀反应器
( ) 7. horizontal grit chamber   g. 污泥干化池；晒渣池
( ) 8. flocculation sedimentation reactor   h. 气浮反应器
( ) 9. sludge drying bed   i. 样板工程
( ) 10. treatment efficiency   j. 均化池

IV. Match the words and expressions with their meanings.

( ) 1. contaminant   a. far from
( ) 2. raw water   b. pollutant
( ) 3. close to   c. purpose
( ) 4. maximum   d. treated water
( ) 5. objective   e. minimum

V. Translate the following sentences into Chinese.

1. The settled and floating materials are removed and the remaining liquid may be discharged to secondary treatment.

2. Secondary treatment may require a separation process to remove the micro-

organisms from the treated water prior to discharge or tertiary treatment.

3. The influent sewage water passes through a bar screen to remove all large objects like cans, rags, sticks, plastic packets etc. carried in the sewage stream.

4. If gross solids are not removed, they become entrained in pipes and moving parts of the treatment plant, and can cause substantial damage and inefficiency in the process.

5. For small sanitary sewer systems, the grit chambers may not be necessary, but grit removal is desirable at larger plants.

VI. Make oral presentation on the process flows of the sewage treatment plant in Changsha Environmental Protection Vocational College.

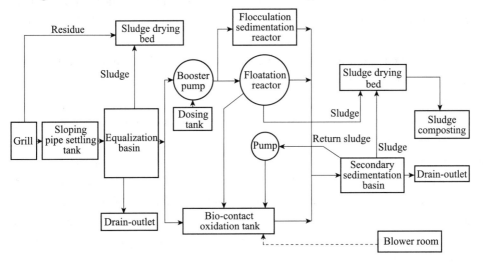

Text B Typical Project of Sewage Treatment Plant (2)

Secondary treatment (Bio-contact oxidation tank/ Flocculation sedimentation reactor/ Floatation reactor):

Secondary treatment is designed to substantially degrade the biological content of the sewage which is derived from human waste, food waste, soaps and detergent. The majority of municipal plants treat the settled sewage liquor using aerobic biological processes.

(1) Bio-contact oxidation is a common process of secondary treatment systems, which can support the growth of bacteria and micro-organisms present in the sewage to break down (分解) and stabilize (抑制) organic pollutants. To be effective, the biota require both oxygen and food to live. The bacteria and protozoa consume biodegradable soluble organic contaminants and bind much of the less soluble fractions into floc. Air blowers in the base of the tank may be used to help provide the dissolved oxygen for aerobic microorganism.

(2) Flocculation sedimentation is widely employed in the purification of drinking water as well as sewage treatment, storm-water treatment and treatment of other industrial wastewater streams.

In Flocculation sedimentation reactor, due to the addition of a clarifying agent, colloids come out of suspension in the form of floc or flake, and settle down to the bottom of the reactor.

The action differs from precipitation in that, prior to flocculation, colloids are merely suspended in a liquid and not actually dissolved in a solution. In the flocculated system, there is no formation of a cake (块状物), since all the flocs are in the suspension.

(3) Dissolved air flotation is a water treatment process that clarifies wastewaters (or other waters) by the removal of suspended matter such as oil or solids. The removal is achieved by dissolving air in the water or wastewater under pressure and then releasing the air at atmospheric pressure in a flotation tank or basin. The released air forms tiny bubbles which adhere to the suspended matter causing the suspended matter to float to the surface of the water where it may then be removed by a skimming device. Dissolved air flotation is very widely used in treating the industrial wastewater effluents from oil refineries, petrochemical and chemical plants, natural gas processing plants and paper mills.

Secondary sedimentation:

The final step in the secondary treatment stage is to settle out the biological floc or filter material through a secondary clarifier and to produce sewage water containing low levels of organic material and suspended matter.

Sludge treatment and disposal:

The sludges accumulated in a wastewater treatment process must be treated and disposed in a safe and effective manner. The purpose of digestion is to reduce the amount of organic matter and the number of disease-causing microorganisms present in the solids. The most common treatment options include anaerobic digestion, aerobic digestion, and composting.

Sludge treatment depends on the amount of solids generated and other site-specific conditions（具体的场地条件）. Composting is most often applied to small-scale plants with aerobic digestion for mid-sized operations, and anaerobic digestion for the larger-scale operations.

Major equipment and structures:

Table1　Major Equipment and Structures

	Device Name	Amount	Specification
1	grill trough	1	2 170×420×600(mm)
2	sloping pipe settling tank	1	5 000×1 100×3 200(mm)
3	wastewater flow equalization basin	1	6 400×3 200×4 800(mm)
4	flocculation sedimentation reactor	1	
5	floatation reactor	1	Dissolved air vessel: φ600×800 Reactor: φ900×800 (Separate part); φ300×3 300 (Reaction part)
6	bio-contact oxidation tank	2	2 600×2 200×5 000(mm)
7	secondary sedimentation basin	1	3 300×1 800×3 600(mm)
8	blower room	1	6 000×6 000(mm)JTS-50 blower×2
9	water pump	8	IHG20-110、IHG25-110、IHG40-125 &32WQ8-12-0.75
10	sludge drying bed	1	7 960×3 000×1 300(mm)

Words and Expressions

pump　[pʌmp] *n.* 泵；抽水机
　　　　　　　　vt. 打气；用抽水机抽……
　　　　　　　　vi. 抽水
substantially　[səb'stænʃəli] *adv.* 实质上；大幅度地；充分地
liquor　['likə] *n.* 液体；溶液

biota　[bai'əutə]　*n.* 生物群（系）；（某一地区、某一时代的）动植物
soluble　['sɒljʊb(ə)l]　*adj.* 可溶解的
bind　[baind]　*vt.* 凝固
clarify　*vt.* 澄清
　　　　vi. 得到澄清；得到净化
form　*n.* 形状；形态
　　　vt. 构成；生产
　　　vi. 构成；形成
specification　[ˌspesifi'keiʃ(ə)n]　*n.* 规格；规范

be derived from　源自于……
the majority of　……的大多数；……的大部分
air blower　鼓风机
clarifying agent　澄清剂
oil refinery　炼油厂
petrochemical chemical plant　石油化工厂
natural gas processing plant　天然气处理厂
paper mill　造纸厂
settle out　沉淀出来；趋于稳定

Notes

1. Bio-contact oxidation is a common process of secondary treatment systems, which can support the growth of bacteria and micro-organisms present in the sewage to break down（分解）and stabilize（抑制）organic pollutants. 生物接触氧化是一种常用的二级水处理工艺，它通过为废水中的细菌和微生物提供生长所需的养分来分解和抑制废水中的有机污染物。
    - process　工艺；流程
    - present in　存在于……；出现于……
    - 本句的主句的基本结构是主系表结构。关系代词 which 引导用逗号跟主句隔开的非限制性定语从句，且在定语从句中作主语，指代先行词 Bio-contact oxidation。
2. The bacteria and protozoa consume biodegradable soluble organic contaminants and bind much of the less soluble fractions into floc. 细菌和原生动物消耗可生物降解的可溶有机污染物，并把相当多的不那么易于溶解的部分物质凝聚成絮状物。
    - soluble　可溶的；可溶解的
3. In Flocculation sedimentation reactor, due to the addition of a clarifying agent, colloids

come out of suspension in the form of floc or flake, and settle down to the bottom of the reactor. 在絮凝沉淀反应器中，由于澄清剂的添加，在絮状和小薄片状的悬浮物中会形成胶粒物质，并会下沉到反应器的底部。
- settle down　沉淀下去；下沉
- in the form of　以……的形式
- floc　絮状物
- flake　小薄片

4. The released air forms tiny bubbles which adhere to the suspended matter causing the suspended matter to float to the surface of the water where it may then be removed by a skimming device. 释放出来的空气形成细小的黏附在悬浮物质上的泡沫，使得这些悬浮物质漂浮到水面，然后就可以被撇沫装置去除。
- adhere to　黏附
- skimming device　撇沫设备；撇油装置；撇具

5. Composting is most often applied to small-scale plants with aerobic digestion for mid-sized operations, and anaerobic digestion for the larger-scale operations. 堆肥法最常用于小规模的污水处理厂，小型的操作采用好氧消化，大规模的操作就采用厌氧消化。
- be applied to　应用于……
  eg. The methods used here can be applied to other projects. 这方法可以应用于其他项目。

Exercises

I. Decide whether each of the following statements is true (T) or false (F) according to the text.

(　) 1. The sewage derived from soaps and detergent contains biological content.
(　) 2. Flocculation sedimentation is widely used in the purification of drinking water.
(　) 3. Effluents paper mills can be called agricultural wastewater.
(　) 4. Primary treatment is to settle out the biological floc or filter material.
(　) 5. The purpose of digestion is to reduce heat waste.

II. Translate the following words or phrases into Chinese.

1. bio-contact oxidation tank

2. clarifying agent

3. flocculation sedimentation reactor

4. grill trough

5. secondary sedimentation basin　　6. blower room

7. equalization basin　　8. settling tank

9. disease-causing microorganisms　　10. filter material

III. Match the words and expressions with their meanings.

(　) 1. sludge drying bed　　a. 堆肥
(　) 2. secondary clarifier　　b. 大气压
(　) 3. precipitation　　c. 絮凝
(　) 4. water pump　　d. 污泥干化池
(　) 5. anaerobic digestion　　e. 悬浮物
(　) 6. floatation reactor　　f. 厌氧消化
(　) 7. suspended matter　　g. 气浮反应器
(　) 8. composting　　h. 水泵
(　) 9. atmospheric pressure　　i. 沉淀
(　) 10. flocculation　　j. 二级澄清池

IV. Match the words and expressions with their meanings.

(　) 1. use　　a. purify
(　) 2. liquid　　b. choice
(　) 3. clarify　　c. employ
(　) 4. option　　d. method
(　) 5. manner　　e. liquor

V. Translate the following sentences into Chinese.

1. The majority of municipal plants treat the settled sewage liquor using aerobic biological processes.

2. Air blowers in the base of the tank may be used to help provide the dissolved oxygen for aerobic microorganism.

3. Dissolved air flotation is a water treatment process that clarifies wastewaters (or other waters) by the removal of suspended matter such as oil or solids.

4. The most common treatment options include anaerobic digestion, aerobic digestion, and composting.

5. Sludge treatment depends on the amount of solids generated and other site-specific conditions.

VI. List the names and amount of the devices in the table in this text.

Part II Air Pollution Control Techniques

Unit 1 Air Pollution Control (Gas)

Text Control of Gas

Air is an important natural resource providing the basis of life on earth. The air in the atmosphere provides oxygen to plants and animals by virtue of which they are able to live. It is therefore important to have good quality air for various activities. However, this is becoming increasingly difficult in view of large scale pollution caused by the industrialization of society, intensification of agriculture, introduction of motorized vehicles and explosion of the population. These activities generate primary and secondary air pollutants which substantially change the composition of air. Kaifu therefore defined air pollution as the introduction of chemicals, particulate matter (PM) or biological materials that cause harm or discomfort to humans or other living organisms, or cause damage to the natural environment or built environment into the atmosphere.

To control gaseous criteria pollutants as well as volatile organic compounds (VOCs) and other gaseous air toxics, absorption, adsorption, and incineration have been identified by ICMA as the three basic techniques employed to control gaseous pollutants. These techniques may be employed singly or in combination depending on the type of pollutant and are detailed below.

Absorption

In the context of air-pollution control, absorption involves the transfer of a gaseous pollutant from the air into a contacting liquid such as water. The liquid must either be able to serve as a solvent for the pollutant or to capture it by means of a chemical reaction. Wet scrubbers similar to those employed to control suspended particulates may be used for gas absorption. Gas absorption can also be carried out in packed scrubbers, or towers, in which the liquid is present on a wetted surface rather than droplets suspended in the air.

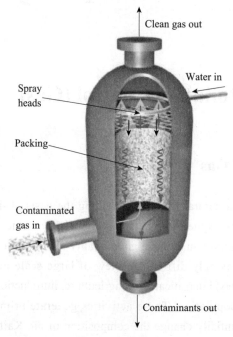

Figure 1　Diagram of packed wet scrubber

Adsorption

Gas adsorption as contrasted with absorption is a surface phenomenon. The gas molecules are sorbed i.e. attracted to and held on the surface of a solid. Gas adsorption methods are used for odour（气味）control at various types and stages of chemical-manufacturing and food-processing facilities in the recovery of a number of volatile solvents (eg., benzene 苯) and in the control of VOCs at industrial facilities.

Activated carbon (heated charcoal) is one of the most common adsorbent materials, which is very porous and has an extremely high ratio of surface area to volume. Activated carbon is particularly useful as an adsorbent for cleaning air-streams that contain VOCs and for solvent recovery and odour control. A properly designed carbon adsorption unit can remove gas with an efficiency exceeding 95%.

Incineration

The process called incineration or combustion can be used to convert VOCs and other gaseous hydrocarbon pollutants to carbon dioxide and water. Incineration of VOCs and hydrocarbon fumes is usually accomplished in a special incinerator called an afterburner. To achieve complete combustion, the afterburner must provide the proper amount of turbulence and burning time and it must maintain a sufficiently high temperature. Sufficient turbulence or mixing, is a key factor in combustion because it reduces the required burning time and temperature. A process called direct flame incineration can be used when the waste gas is itself a combustible mixture and does not need the addition of air or fuel.

Words and Expressions

industrialization [in,dʌstriəlai'zeiʃən] n. 工业化
intensification [in,tensifi'keiʃən] n. 集约化
substantially [səb'stænʃəli] adv. 极大地；相当多地
composition [kɒmpə'ziʃ(ə)n] n. 成分；构成
solvent ['sɔlvənt] n. 溶剂
droplet ['drɔplit] n. 小水滴
suspend [sə'spend] v. 悬浮
absorption [əb'zɔːpʃ(ə)n] n. 吸收
adsorption [æd'sɔːpʃən] n. 吸附
incineration [in,sinə'reiʃn] n. 焚化
porous ['pɔːrəs] adj. 有气孔的；能渗透的
volume ['vɒljuːm] v. 把……收集成卷
combustion [kəm'bʌstʃ(ə)n] n. 燃烧；烧毁
incinerator [in'sinəreitə] n.（废物的）焚化炉
afterburner ['ɑːftə,bɜːnə] n. 加力燃烧室；后燃室

primary air pollutant 初次污染物
secondary air pollutant 二次污染物

particulate matter 颗粒污染物
living organism 生物有机体
gaseous criteria pollutant 气体标准污染物
volatile organic compounds (VOCs) 挥发性有机化合物
gaseous air toxics 气态有毒物
contacting liquid 接触液体
packed scrubber 填料洗涤器；填料除尘器
wet scrubber 湿式除尘器
volatile solvent 挥发性溶剂
activated carbon 活性炭
gaseous hydrocarbon pollutant 气态烃污染物
carbon dioxide 二氧化碳
hydrocarbon fume 烃废气；油烟气
complete combustion 完全燃烧
direct flame incineration 火焰直接焚烧
combustible mixture 可燃混合物

Notes

1. Kaifu therefore defined air pollution as the introduction of chemicals, particulate matter (PM) or biological materials that cause harm or discomfort to humans or other living organisms, or cause damage to the natural environment or built environment into the atmosphere. 因此，Kaifu 将空气污染定义为由化学物、颗粒污染物（PM）或者生物材料的引入而给人类或者其他生物有机体造成伤害或不适，或者给自然环境或建筑环境造成损害。
   - define…as… 将……定义为……
     eg. We define life as a challenge. 我们将生命定义为一种挑战。

2. To control gaseous criteria pollutants as well as volatile organic compounds (VOCs) and other gaseous air toxics, absorption, adsorption, and incineration have been identified by ICMA as the three basic techniques employed to control gaseous pollutants. 为了控制气态标准污染物、挥发性有机化合物 (VOC) 及其他气态有毒物，ICMA 已确定通过利用吸收、吸附、焚烧三种基本技术来控制气态污染物。
   - be identified as… 被认出；被确定为……
     eg. She identified the man as her attacker. 她认出那个男人就是袭击过她的人。
     Some of them are identified as priority projects in China's 11th Five-Year Plan for Economic and Social Development. 其中一些被确定为中国"十一五"经济和社会

发展计划的优先项目。

3. Gas absorption can also be carried out in packed scrubbers, or towers, in which the liquid is present on a wetted surface rather than droplets suspended in the air. 这种气体吸收也可以在填料除尘器或塔内进行，其中的液体存在于一个潮湿的表面上，而不是像水滴一样悬浮在空中。
   - carry out 实施；进行
   - wetted surface 潮湿的表面
   - droplets suspended in the air 悬浮于空中的小水滴

4. Gas adsorption methods are used for odour control at various types and stages of chemical-manufacturing and food-processing facilities in the recovery of a number of volatile solvents (eg. benzene) and in the control of VOCs at industrial facilities. 气体吸附方法用于在一些挥发性溶剂（如苯）的恢复和工业挥发性有机化合物的控制过程中对各类型和阶段的化工生产和食品加工进行气味控制。
   - chemical-manufacturing 化工生产
   - food-processing 食品加工

5. Activated carbon (heated charcoal) is one of the most common adsorbent materials, it is very porous and has an extremely high ratio of surface area to volume. 活性炭（加热木炭）是最常用的吸附材料之一，这种材料多孔，具有极高的表面吸附率。
   - activated carbon 活性炭
   - heated carbon 热成炭
   - adsorbent materials 吸附材料
   - an extremely high ratio of surface area to volume 极高的表面吸附率
   - volume 该词原意为"把……收集成卷"，这里是指"活性炭对颗粒的吸附"

6. A process called direct flame incineration can be used when the waste gas is itself a combustible mixture and does not need the addition of air or fuel. 当废气本身就是一种可燃混合物且不需要添加空气或燃料时，便可使用所谓的"火焰直接焚烧法"。
   - the addition of air or fuel 空气或燃料的添加

Exercises

I. Answer the following questions after reading the text.

1. Why is it becoming increasingly difficult to have good air quality nowadays?
2. How many types of pollutants do people's activities generate? What are they?
3. What's air pollution according to Kaifu?
4. What are the three basic techniques employed to control gaseous pollutants?

5. What devices can be used for gas absorption?

II. Decide whether each of the following statements is true (T) or false (F) according to the text.

(　) 1. In the process of absorption, contacting liquid such as water must either be able to serve as a solvent for the pollutant or to capture it by means of a chemical reaction.

(　) 2. When gas absorption is carried out in packed scrubbers or towers, the liquid in those devices is present as droplets suspended in the air rather than on a wetted surface.

(　) 3. Absorption is a surface phenomenon.

(　) 4. The process of incineration or combustion can generate carbon dioxide and some less polluting air.

(　) 5. Turbulence is necessary for complete combustion.

III. Translate the following words or phrases into English.

1. 初次污染物　　　　　2. 二次污染物

3. 颗粒污染物　　　　　4. 气体标准污染物

5. 填料除尘器　　　　　6. 活性炭

7. 后燃室　　　　　　　8. 挥发性有机化合物

9. 气味控制　　　　　　10. 喷头

IV. Translate the following sentences into Chinese.

1. However, this is becoming increasingly difficult in view of large scale pollution caused by the industrialization of society, intensification of agriculture, introduction of motorized vehicles and explosion of the population.

2. In the context of air-pollution control, absorption involves the transfer of a gaseous pollutant from the air into a contacting liquid such as water.

3. Activated carbon is particularly useful as an adsorbent for cleaning air-streams that contain VOCs and for solvent recovery and odour control.

4. The process called incineration or combustion can be used to convert VOCs and other gaseous hydrocarbon pollutants to carbon dioxide and water.

5. To achieve complete combustion, the afterburner must provide the proper amount of turbulence and burning time and it must maintain a sufficiently high temperature.

V. Match the words and expressions with their meanings.

1. living organism   a. 气态污染物
2. Absorption   b. 可燃混合物
3. gaseous pollutant   c. 烃废气
4. Adsorption   d. 吸收
5. volatile solvent   e. 污染气体
6. Incineration   f. 火焰直接焚烧
7. hydrocarbon fumes   g. 生物有机体
8. direct flame incineration   h. 挥发性溶剂
9. combustible mixture   i. 焚烧
10. contaminated gas   j. 吸附

VI. Best choice.

1. Absorption involves the transfer of a gaseous pollutant from the air into _____.
A. solid   B. air   C. contacting liquid   D. carbon dioxide

2. What is one of the most common adsorbent materials mentioned in the text?
A. contacting liquid such as water   B. clean air
C. activated carbon   D. VOCs

3. Why does activated carbon have an extremely high ratio of surface area to volume?
    Because it is _____.
A. sticky   B. porous   C. magnetic   D. attractive

4. Activated carbon is particularly useful as an adsorbent for _____.
A. solvent recovery and odour control
B. transferring dirty air into water
C. cleaning air-streams that contain particulate matter
D. complete combustion

5. Where is incineration of VOCs and hydrocarbon fumes usually accomplished?
   A. in a packed tower   B. in a packed scrubber
   C. in a carbon adsorption unit D. in an afterburner

VII. Fill in the following blanks.

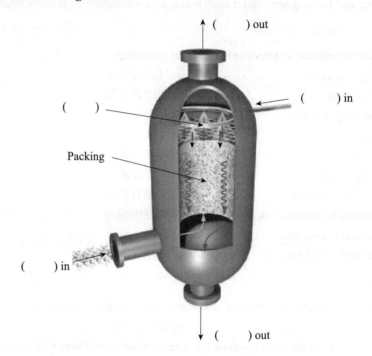

Unit 2　Air Pollution Control (Particulates)

Text　Control of Particulates

Airborne particles（尘粒）can be removed from a polluted airstream by a variety of physical processes. The common types of equipment for collecting—fine particulates include cyclones, scrubbers, electrostatic precipitators and bag house filters. Once collected, particulates adhere to each other forming agglomerates（成团）that can readily be removed from the equipment and disposed off usually in a landfill. Electrostatic precipitators and fabric-filter bag houses are often used at power plants and the principles behind cyclones, scrubbers, electrostatic precipitators and bag houses as air-cleaning equipment described by ICMA are presented below.

Cyclones

A cyclone (see Figure 1) removes particulates by causing the dirty airstream to flow in a spiral（螺旋）path inside a cylindrical chamber. Dirty air enters the chamber from a tangential（切线的；外围的）direction at the outer wall of the device forming a vortex （旋涡）as it swirls（打旋）within the chamber. The larger particulates because of their greater inertia（惯性）move outward and are forced against the chamber wall. Slowed by friction with the wall surface, they then slide down the wall into a conical（圆锥形的）dust hopper（漏斗）at the bottom of the cyclone. The cleaned air swirls upward in a narrower spiral through an inner cylinder（圆筒）and emerges from an outlet（出口）at the top and accumulated particulate dust is periodically removed from the hopper for disposal. Cyclones are best at removing relatively coarse particulates. They can routinely achieve efficiencies of 90 percent for particles larger than about 20 μm (0.0008 inch). However, cyclones are not sufficient to meet stringent air quality standards. They are typically used as precleaners and are followed by more efficient air- cleaning equipment such as electrostatic precipitators and bag houses.

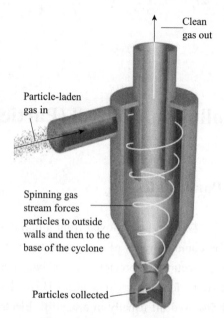

Figure 1 Diagram of a cyclone

Wet scrubbers

Wet scrubbers trap（捕获）suspended particles by direct contact with a spray of water or other liquid. In effect, a scrubber（洗涤器）washes the particulates out of the dirty airstream as they collide with and are entrained by the countless tiny droplets in the spray.

Electrostatic precipitators

Electrostatic precipitation is a commonly used method for removing fine particulates from airstreams. In an electrostatic precipitator (see Figure 2), particles suspended in the airstream are given an electric charge as they enter the unit（装置）and are then removed by the influence of an electric field. The precipitation unit comprises baffles（隔板）for distributing airflow, discharge and collection electrodes, a dust clean-out system and collection hoppers. A high DC voltage（电压）(as much as 100,000 volts) is applied to the discharge electrodes to charge the particles, which then are attracted to oppositely charged collection electrodes on which they become trapped.

In a typical unit, the collection electrodes comprise a group of large rectangular（矩形的）metal plates suspended vertically and parallel to each other inside a boxlike structure.

There are often hundreds of plates having a combined surface area of tens of thousands of square metres. Rows of discharge electrode wires hang between the collection plates. The wires are given a negative electric charge whereas the plates are grounded and thus become positively charged.

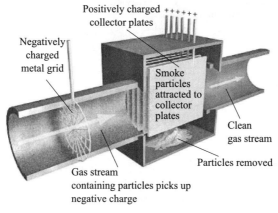

Figure 2　Diagram of an electrostatic precipitator

Bag house filters

One of the most efficient devices for removing suspended particulates is an assembly of fabric filter bags commonly called a bag house. A typical bag house (see Figure 3 below) comprises an array(列) of long, narrow bags—each about 25 cm (10 inches) in diameter (直径) —that are suspended upside down in a large enclosure. Dust-laden air is blown upward through the bottom of the enclosure by fans. Particulates are trapped inside the filter bags while the clean air passes through the fabric and exits at the top of the bag house.

All the particulate control technologies discussed here has its own advantages and disadvantages. ESPs can handle very large volumetric flow rates at low pressure drops and can achieve very high efficiencies (99.9%). They are roughly equivalent in costs to fabric filters and are relatively inflexible to changes in process operating conditions. Wet scrubbers can also achieve high efficiencies and have the major advantage that some gaseous pollutants can be removed simultaneously with the particulates. However, they can only handle smaller gas flows (up to 3,000 m^3/min), can be very costly to operate (owing to a high pressure drop), and produce a wet sludge that can present disposal problems. For a higher flue gas flow rate and greater than 99% removal of PM, ESPs and fabric filters are the equipment of choice, with very little difference in costs.

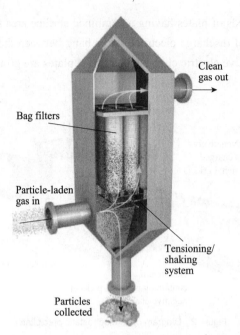

Figure 3　Diagram of typical bag filter (Adapted from DEHP, 2011)

Recommendations

For effective PM_{10} control in industrial application, the use of ESPs or baghouses is recommended. They should be operated at their design efficiencies. In the absence of a specific emissions requirement, a maximum level of 50 milligrams per normal cubic meter (mg/Nm^3) should be achieved.

For gases containing soluble toxics and where the gas flow rate is less than 3,000 m^3/min, wet scrubbers may be used. Cyclones and mechanical separators should be used only as precleaning devices upstream of a baghouse or an ESP.

Words and Expressions

particulate　[pɑː'tikjʊlət]　*n.* 微粒
airstream　['eəstriːm]　*n.* 气流
cyclone　['saikləʊn]　*n.* 旋风
scrubber　['skrʌbə]　*n.* 洗涤器；除尘器
electrostatic　[i,lektrə(ʊ)'stætik]　*n.* 静电的

precipitator　[pri'sipiteitə]　*n.* 沉淀器
filter　['filtə]　*n.* 过滤器
adhere　[əd'hiə]　*n.* 黏附；附着
landfill　['læn(d)fil]　*n.* 垃圾填埋场
slide　[slaid]　*n.* 滑落；下跌
outlet　['aʊtlet]　*n.* 出口；通风口
periodically　[,piəri'ɒdikəli]　*adv.* 周期性地
relatively　['relətivli]　*adv.* 相对地；比较地
coarse　[kɔːs]　*adj.* 粗糙的
routinely　[ruː'tiːnli]　*adv.* 常规地；惯常地
precleaner　[priː'kliːnə]　*n.* 预滤器
trap　[træp]　*n.* 捕获；困住
spray　[sprei]　*n.* 喷雾
collide　[kə'laid]　*vt. & vi.* 相撞；碰撞
entrain　[in'trein]　*v.* 拖；产生
comprise　[kəm'praiz]　*vt.* 包括；由……组成
airflow　['eəfləʊ]　*n.* 气流
electrode　[i'lektrəʊd]　*n.* 电极
charge　[tʃɑːdʒ]　*vi.* 充电
particle　['pɑːtik(ə)l]　*n.* 微粒，颗粒
vertically　['vəːtikəli]　*adv.* 垂直地
parallel　['pærəlel]　*adj.* 平行的
negative　['negətiv]　*adj.* 负的
positive　['pɒzitiv]　*adv.* 正的
enclosure　[in'kləʊʒə]　*n.* 圈占空间；封起来的空间
recommend　[rekə'mend]　*vt.* 推荐；介绍
　　　　　　　　　　　　　　　vi. 推荐；建议

airborne particles　尘粒
fine particulates　细粒子
electrostatic precipitator　静电除尘器
bag house filter　袋式除尘器
fabric-filter bag　织物过滤袋
power plant　发电厂
cylindrical chamber　圆柱腔

conical dust hopper 锥形尘斗
suspended particle 悬浮颗粒物
electric charge 电荷
electric field 电场
collection hopper 收集漏斗
DC voltage 直流电压
rectangular metal plate 矩形金属板
dust-laden air 有尘空气；含尘空气
industrial application 工业应用

Notes

1. Once collected, particulates adhere to each other forming agglomerates that can readily be removed from the equipment and disposed off usually in a landfill. 颗粒物一旦收集起来，它们便彼此黏附在一起成团，以便随时可以从设备中清除或者将其在垃圾填埋场处理掉。
   - adhere to 黏附；附着
     eg. Paste is used to make one surface adhere to another. 糨糊是用以使一个接触面黏住另一个接触面的。
     This paint will adhere to any surface, whether rough or smooth. 这种油漆能附着于任何粗糙或光滑的表面。
   - dispose off 处理掉
     eg. What is the safest means of disposing off nuclear waste? 处理核废料最安全的方法是什么？

2. Dirty air enters the chamber from a tangential direction at the outer wall of the device forming a vortex as it swirls within the chamber. 脏空气从旋风分离器的外壁沿切线方向进入圆柱腔，在圆柱腔内打旋的脏气流形成旋涡，较大的颗粒物由于惯性较大而向外侧移动，并被迫旋转至圆柱腔壁。
   - from a tangential direction 沿切线方向
   - forming a vortex 形成一个旋涡
   - swirls within the chamber 在圆柱腔内打旋

3. The cleaned air swirls upward in a narrower spiral through an inner cylinder and emerges from an outlet at the top and accumulated particulate dust is periodically removed from the hopper for disposal. 清洁的空气在内腔里呈细细的螺旋线向上旋转，并从顶部的出口涌出，聚集的颗粒粉尘将从漏斗中被定期地移除并得到处置。
   - inner cylinder 内筒；内腔

Part II　Air Pollution Control Techniques　103

- emerge from 从……中出现
  eg. Two reasons emerge from the interviews. 采访中发现两个方面的原因。
  There is growing evidence that the economy is at last emerging from recession. 有越来越多的证据表明经济终于开始摆脱萧条了。

4. In an electrostatic precipitator (see Figure 2), particles suspended in the airstream are given an electric charge as they enter the unit and are then removed by the influence of an electric field. 在一个静电除尘器中（图 2），脏气流中的悬浮颗粒在进入该装置时便携带电荷并在电场的作用下被清除。
- unit 装置；装备。这里是指上文提到的静电除尘器

5. A high DC voltage (as much as 100,000 volts) is applied to the discharge electrodes to charge the particles, which then are attracted to oppositely charged collection electrodes on which they become trapped. 高直流电压（高达 100,000 V）作用于放电电极以给颗粒充电，带电颗粒即被吸引至相反电荷收集电极而被捕获住。
- DC voltage　直流电压
- be applied to doing something. 运用；被运用于……
  此处的"to"作介词，后接名词或名词词组。
  eg. We should apply what we have learnt to our practical work. 我们应该把我们所学的应用于我们的实际工作中。
  Three techniques might be applied to the control of gases. 三种技术可运用于气体控制。
- charge the particles　给粒子充电
- oppositely charged collection electrodes　被充上正电的收集电极

6. In a typical unit, the collection electrodes comprise a group of large rectangular metal plates suspended vertically and parallel to each other inside a boxlike structure. 在一个典型的静电除尘装置里，汇集电极由一组大型的矩形金属板组成，这些金属板垂直地分布在一个结构箱里并相互平行。
- comprise v. 包含；包括；由……组成 / 构成
  eg. The country comprises 20 states. 这个国家包含 20 个州。
  这个句式就是"整体 +comprises(一定要用单数，主语是整体) + 部分"。
  Twelve departments comprise this university. 12 个系组成了这所大学。
  这时的用法就是"部分 +comprise（要用复数，主语是很多个部分）+ 整体"。
- suspended vertically and parallel to each other　其中的 vertically 是相对于结构箱底部而言，而 parallel 则是相对于"矩形金属板之间"而言的，因此可译为：这些金属板垂直地分布在一个结构箱里并相互平行。

7. In the absence of a specific emissions requirement, a maximum level of 50 milligrams

per normal cubic meter (mg/Nm^3) should be achieved. 如果没有具体的排放要求，应要达到小于 50mg/m^3 的排放量。

- in the absence of 不存在；缺少……时；无……时

Exercises

I. Answer the following questions after reading the text.

1. What are the common types of equipment for collecting fine particulates?
2. What equipments are often used at power plants?
3. Where do the cleaned air and accumulated particulate come out respectively in a cyclone?
4. Why can the particles be trapped on the collection electrodes in an electrostatic precipitator?
5. What does a precipitation unit comprise?

II. Decide whether each of the following statements is true (T) or false (F) according to the text.

( ) 1. All the equipment remove airborne particles from a polluted airstream by a variety of chemical processes.

( ) 2. A cyclone removes particulates by causing dirty airstream to flow in a straight path in a cylindrical chamber.

( ) 3. Cyclones are not sufficient to meet stringent air quality standards, whereas electrostatic precipitators and bag houses are more efficient in cleaning air.

( ) 4. In electrostatic precipitator, the rectangular metal plates suspended vertically to each other inside a boxlike structure.

( ) 5. Bag house filters are often used for collecting coarse particulates.

III. Translate the following words or phrases into English.

1. 悬浮颗粒 _____ 2. 旋风 _____

3. 静电除尘器 _____ 4. 湿式除尘器 _____

5. 袋式除尘器 _____ 6. 电荷 _____

7. 电场 _____ 8. 直流电压 _____

9. 织物过滤袋　　　　　　　　10. 锥形尘斗

IV. Translate the following sentences into Chinese.

1. A cyclone removes particulates by causing the dirty airstream to flow in a spiral path inside a cylindrical chamber.

2. In effect, a scrubber washes the particulates out of the dirty airstream as they collide with and are entrained by the countless tiny droplets in the spray.

3. One of the most efficient devices for removing suspended particulates is an assembly of fabric filter bags commonly called a bag house.

4. A typical bag house comprises an array of long, narrow bags—each about 25 cm (10 inches) in diameter—that are suspended upside down in a large enclosure.

V. Match the words and expressions with their meanings.

1. airborne particles　　　　　　　a. 正电集流板
2. particle-laden gas　　　　　　　b. 严格的空气质量标准
3. positively charged collector plates　c. 脏气流
4. negatively charged metal grid　　d. 细粒子
5. landfill　　　　　　　　　　　　e. 尘粒
6. stringent air quality standards　　f. 含尘空气
7. dirty airstream　　　　　　　　g. 排放电极
8. fine particulates　　　　　　　　h. 负电金属网格
9. discharge electrodes　　　　　　i. 收集漏斗
10. collection hopper　　　　　　　j. 垃圾填埋场

VI. Best choice.

1. The larger particulates are forced against the chamber wall because of _____.

A. the attraction of the wall

B. the fast speed of the spinning gas

C. the greater inertia of the particulates

D. large space of the chamber

2. What causes the large particulates to slow down in the chamber and slide down the wall?

A. friction　　　　B. inertia　　　　C. the cleaned air　　　　D. the spanning gas

3. Which of the following equipment washes the particulates out of the dirty airstream by the countless tiny droplets in the spray?

A. electrostatic precipitators   B. cyclones

C. wet scrubbers   D. bag house filters

4. _____ is necessary for an electrostatic precipitator to charge the particles?

A. A high DC voltage   B. A high AC voltage

C. A liquid such as water   D. A fan

5. Where are the particulates trapped in a bag house filter?

A. on the surface of the bag filters   B. on the wall

C. at the bottom of the enclosure   D. inside the bag filters

VII. Fill in the following blanks.

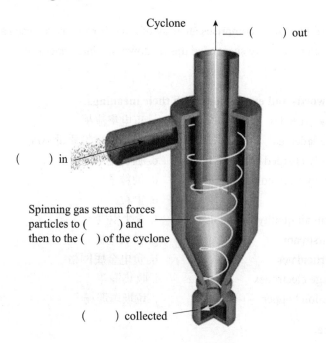

Cyclone

——( ) out

( ) in

Spinning gas stream forces particles to ( ) and then to the ( ) of the cyclone

( ) collected

Part II　Air Pollution Control Techniques　107

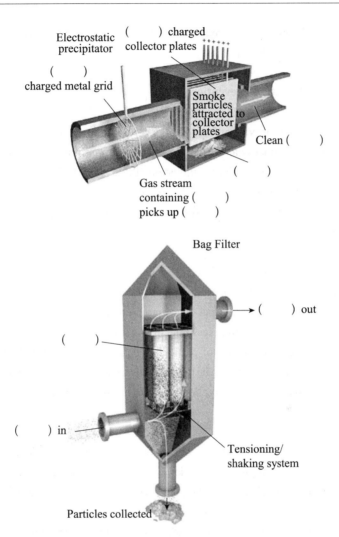

VIII. Work in pairs or groups and list the principal advantages and disadvantages of the particulate control technologies discussed in this unit.

IX. Reading Material

Airborne Particulate Matter: Pollution Prevention and Control

　　Airborne particulate matter (PM) (颗粒物) emissions can be minimized by pollution prevention and emission control measures. Prevention, which is frequently more cost-effective than control, should be emphasized. Special attention should be given to pollution abatement measures in areas where toxics associated with particulate emissions may pose a significant environmental risk.

Approaches to Pollution Prevention

Management

Measures such as improved process design, operation, maintenance, housekeeping, and other management practices can reduce emissions. By improving combustion efficiency, the amount of products of incomplete combustion (PICs 不完全燃烧产物), a component of particulate matter, can be significantly reduced. Proper fuel-firing practices and combustion zone configuration（燃烧区域配置）, along with an adequate amount of excess air, can achieve lower PICs.

Choice of Fuel

Atmospheric particulate emissions can be reduced by choosing cleaner fuels. Natural gas used as fuel emits negligible amounts of particulate matter. Oil-based processes also emit significantly fewer particulates than coal-fired combustion processes. Low-ash（低灰的）fossil fuels（化石燃料）contain less noncombustible（不燃的）, ash-forming mineral matter and thus generate lower levels of particulate emissions. Lighter distillate（蒸馏液；馏分油）oil-based combustion results in lower levels of particulate emissions than heavier residual（残留的）oils. However, the choice of fuel is usually influenced by economic as well as environmental considerations.

Fuel Cleaning

Reduction of ash by fuel cleaning reduces the generation of PM emissions. Physical cleaning of coal through washing and beneficiation（选矿）can reduce its ash and sulfur（硫）content, provided that care is taken in handling the large quantities of solid and liquid wastes that are generated by the cleaning process. An alternative to coal cleaning is the co-firing（掺烧）of coal with higher and lower ash content. In addition to reduced particulate emissions, low-ash coal also contributes to better boiler（锅炉）performance and reduced boiler maintenance costs and downtime, thereby recovering some of the coal cleaning costs. For example, for a project in East Asia, investment in coal cleaning had an internal rate of return of 26%.

Choice of Technology and Processes

The use of more efficient technologies or process changes can reduce PIC emissions. Advanced coal combustion technologies such as coal gasification（煤炭气化）and fluidized-bed combustion（流化床燃烧）are examples of cleaner processes that may lower PICs by approximately 10%. Enclosed coal crushers and grinders emit lower PM.

Approaches to Emission Control

A variety of particulate removal technologies, with different physical and economic characteristics, are available.

Inertial or impingement separators（撞击式分离器）rely on the inertial properties of the particles to separate them from the carrier gas stream（载气流）. Inertial separators are primarily used for the collection of medium-size and coarse particles. They include settling chambers（沉降室）and centrifugal cyclones（离心旋风机）(straight-through, or the more frequently used reverse-flow cyclones)（直流或更常用的反流旋风机）.

Electrostatic precipitators (*ESPs*) remove particles by using an electrostatic field（静电场）to attract the particles onto the electrodes.

Filters and dust collectors (*baghouses*) collect dust by passing flue gases through a fabric that acts as a filter. The most commonly used is the bag filter, or baghouse. The various types of filter media include woven fabric（织物）, needled felt（针刺毛毡）, plastic, ceramic（陶瓷）, and metal.

Wet scrubbers rely on a liquid spray to remove dust particles from a gas stream. They are primarily used to remove gaseous emissions, with particulate control a secondary function.

Equipment Selection

The selection of PM emissions control equipment is influenced by environmental, economic, and engineering factors:

Environmental factors include (a) the impact of control technology on ambient air quality; (b) the contribution of the pollution control system to the volume and characteristics of wastewater and solid waste generation; and (c) the maximum which allows able emissions requirements.

Economic factors include (a) the capital cost of the control technology; (b) the operating and maintenance costs of the technology; and (c) the expected lifetime and salvage value of the equipment.

Engineering factors include (a) contaminant characteristics such as physical and chemical properties（性质）—concentration, particulate shape, size distribution, chemical reactivity, corrosivity, abrasiveness, and toxicity; (b) gas stream characteristics such as volume flow rate, dust loading, temperature, pressure, humidity, composition, viscosity, density, reactivity, combustibility, corrosivity, and toxicity; and (c) design and performance characteristics of the control system such as pressure drop, reliability, dependability, compliance with utility and maintenance requirements, and temperature limitations, as well as size, weight, and fractional efficiency curves for particulates and mass transfer or contaminant destruction capability for gases or vapors.

Decide whether each of the following statements is true (T) or false (F) according to the given materials.

( ) 1. Control, which is frequently more cost-effective than Prevention, should be emphasized.
( ) 2. The amount of products of incomplete combustion can be significantly reduced by improving combustion efficiency.
( ) 3. Lower PICs can be achieved only by an adequate amount of excess air.
( ) 4. Coal-fired combustion processes emit fewer particulates than Oil-based processes.
( ) 5. Low-ash fossil fuels generate lower levels of particulate emissions because they contain less noncombustible, ash-forming mineral matter.
( ) 6. Reduction of ash by fuel cleaning reduces the generation of PM emissions.
( ) 7. Low-ash coal contributes to reduced particulate emissions.
( ) 8. Inertial separators are primarily used for the collection of fine particulates.
( ) 9. Electrostatic precipitators remove particles by using an electrostatic field to attract the particles onto the electrodes.
( ) 10. Three factors such as environmental, economic, and engineering factors can influence the selection of PM emissions control equipment.

Part III Solid Waste Management

Unit 1 Introduction to Solid Waste Management

Text Solid Waste

Humans have always produced trash and have always disposed of it in some way, so solid waste management is not a new issue. What have changed are the types and amounts of waste produced, the methods of disposal, and the human values and perceptions of what should be done with it.

In the past, refuse（废物；垃圾）was typically discarded in the most convenient manner possible with little regard to its effects on human health or the environment. Before modern notions of hygiene developed, city streets were typically open sewers（排水沟；下水道）that bred（繁殖；孕育）diseases such as cholera（霍乱）and dysentery（痢疾）. Even until the middle of the twentieth century, household trash was commonly disposed of and burned in open dumps that were neighborhood eyesores（难看的东西；眼中钉）, emitted offensive odors, and attracted rats and other vermin（害虫；害兽）. Chemical wastes were often haphazardly stored in on-site industrial piles or treatment ponds. Particularly noxious waste might be buried, but few controls existed to keep the toxic substances in them from seeping into nearby surface water or contaminating groundwater.

The Basics of Solid Waste Management

Although the terms solid waste, refuse, garbage （垃圾）, and trash are often used interchangeably （可交换地）, solid waste professionals （专业人员）distinguish between them. Solid waste and refuse are synonyms （同义词） that refer to any of a variety of materials that are rejected or discarded as useless.

The variety of materials referred to as solid waste or refuse is broken into several categories:

Garbage （垃圾） strictly refers to animal or vegetable wastes, particularly by-products of food preparation. Garbage decomposes rapidly if exposed to the elements （自然环境） and creates offensive odors.

Trash refers to solid waste that does not decompose (eg., packaging, bottles, cans, building materials).

Hazardous waste （有害垃圾；危险废物） refers to waste that is ignitable, corrosive, or reactive (explosive) or that contains certain concentrations of toxic chemicals specified by the U.S. Environmental Protection Agency (EPA). In addition, the EPA maintains a list of about 500 other specific waste types also considered hazardous. (Although federal laws strictly regulate the generation, transport, and disposal of hazardous waste at most industrial and commercial facilities, hazardous materials in household trash remain exempt from these laws.)

Since most people do not make distinctions between the terms garbage and trash in everyday language, these terms are used interchangeably throughout this background.

Contents of the Solid Waste Stream

Most people do not spend time wondering about what types of materials they throw away or what exactly comprises （包含；由……组成） a garbage truck's contents. But if you were to ask someone what category of material might make up the biggest portion of the truck's contents, you would probably get many different responses. Perceptions of the makeup, or characterization, of the solid waste stream are affected by many factors, including personal consumption, media reports, and visual impressions of litter and overflowing trash cans. The EPA and other government agencies periodically compile data on the contents of our national municipal solid waste (MSW) stream. The figure below summarizes key information from a 1996 EPA report that provides data about the characterization of U.S. MSW broken down by products and materials.

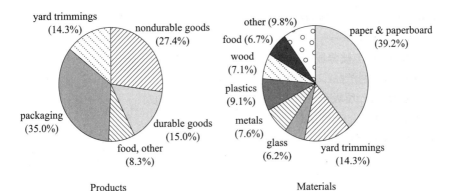

Figure 1 The contents of municipal solid waste by products and materials

The MSW characterized （描述） in the EPA report includes waste from residential, commercial, institutional, and industrial sources. (Industrial waste here includes only packaging and administrative waste, not hazardous or process waste. Other kinds of solid waste, such as agricultural waste and municipal sludge, are not addressed in the EPA report.)

Words and Expressions

trash　[træʃ] *n.* 废物；垃圾
discard　[di'skɑːd] *vt.* 抛弃；放弃；丢弃
　　　　　　　　　　 vi. 放弃
　　　　　　　　　　 n. 抛弃；被丢弃的东西或人
emit　[i'mit] *vt.* 排放；发出
haphazardly　[ˌhæp'hæzədli] *adv.* 随意地；杂乱地
noxious　['nɒkʃəs] *adj.* 有害的；有毒的
decompose　[diːkəm'pəʊz] *vt. vi.* 分解；腐烂
reject　[ri'dʒekt] *vt.* 丢弃；拒绝；排斥；抵制；
hazardous　['hæzədəs] *adj.* 危险的

solid waste management　固体废弃物管理
with little regard to　几乎不考虑
open dump　露天垃圾场
offensive odor　恶臭
on-site　现场的

be referred to as 被称为……
refer to 指的是……
be broken into 被分成……
by-product 副产品
garbage truck 垃圾车
key information 关键信息
be addressed 被提到

Notes

1. What have changed are the types and amounts of waste produced, the methods of disposal, and the human values and perceptions of what should be done with it. 改变了的只是产生的废弃物的类型和数量、处理废弃物的方法、人们对该怎样处理废弃物的价值观念。
   - values 价值观；世界观
   - perceptions 观念
   - do with 处理；处置；对待；运用（一般与 what 连用）
     eg. What should I do with this old computer? 这台旧电脑怎么处理？
2. Particularly noxious waste might be buried, but few controls existed to keep the toxic substances in them from seeping into nearby surface water or contaminating groundwater. 尤其是有毒废弃物可能会被直接填埋，却几乎未采取任何治理措施来阻止废弃物中的有毒物质渗入到附近的地表水中或污染地下水。
   - existed 存在的
   - keep from 阻止；使免于
     eg. Keep away from heat. 请勿靠近热源。
     He said this would prevent companies from creating new jobs. 他说这将阻止各公司创造新的就职岗位。
   - contaminate 污染；弄脏
     eg. Have any fish been contaminated in the Arctic Ocean? 北冰洋里有鱼受到污染了吗？
3. Hazardous waste refers to waste that is ignitable, corrosive, or reactive (explosive) or that contains certain concentrations of toxic chemicals specified by the U.S. Environmental Protection Agency (EPA). 有害废弃物是指具有可燃性、腐蚀性或活性（爆炸性）的废弃物，以及那些含有一定浓度的某些由美国环境保护局认定的有毒化学制品的废弃物。
4. Although federal laws strictly regulate the generation, transport, and disposal of

hazardous waste at most industrial and commercial facilities, hazardous materials in household trash remain exempt from these laws. 虽然联邦法律对大多数工业和商业设施中有害废物的产生、搬运和去除有严格的规定，却不追究家庭生活垃圾中的有害物质的法律责任。
- federal law 联邦法律
- household trash 家庭生活垃圾
- exempt from 豁免；免除
- remain exempt from these laws 可豁免于这些法律

5. Perceptions of the makeup, or characterization, of the solid waste stream are affected by many factors, including personal consumption, media reports, and visual impressions of litter and overflowing trash cans. 人们对于固体废物流的构成或特征的理解受到很多因素的影响，这些因素包括个人消费、媒体报道以及对乱丢的垃圾和满溢的垃圾桶的直观印象。

6. The figure below summarizes key information from a 1996 EPA report that provides data about the characterization of U.S. MSW broken down by products and materials. 下图总结了1996年美国环境保护局报告里的重要信息，该报告提供了根据产品和材料分类描述的美国城市固体废弃物垃圾特性的数据。
- characterization 描述；特性描述
- break down （对观点、陈述等）分门别类
    eg. The report breaks down the results region by region. 该报告将结果按地区一一分类。

Exercises

I. Translate the following words or phrases into English.

1. 固体废弃物 _____
2. 类型 _____
3. 数量；总额 _____
4. 垃圾 _____
5. 产品 _____
6. 分解；腐烂 _____
7. 渗入；渗漏 _____
8. 下水道 _____
9. 有毒的 _____
10. 乱丢 _____

II. Match the words and expressions with their meanings.

( ) 1. household trash   a. 垃圾车
( ) 2. human health   b. 美国环境保护局
( ) 3. offensive odor   c. 城市固体废弃物
( ) 4. garbage truck   d. 垃圾桶
( ) 5. EPA   e. 发出；散发
( ) 6. MSW   f. 包装
( ) 7. trash can   g. 恶臭
( ) 8. hazardous waste   h. 家庭生活垃圾
( ) 9. packaging   i. 人体健康
( ) 10. emit; emission   j. 有毒废物

III. Match the words and expressions with their meanings.

( ) 1. noxious   a. problem
( ) 2. issue   b. waste
( ) 3. refuse   c. toxic
( ) 4. element   d. make up of
( ) 5. contaminate   e. reject
( ) 6. discard   f. dangerous
( ) 7. comprise   g. pollute
( ) 8. hazardous   h. natural environment

IV. Translate the following sentences into Chinese.

1. Humans have always produced trash and have always disposed of it in some way.

2. In the past, refuse was typically discarded in the most convenient manner possible with little regard to its effects on human health or the environment.

3. Solid waste and refuse are synonyms that refer to any of a variety of materials that are rejected or discarded as useless.

4. Trash refers to solid waste that does not decompose (eg., packaging, bottles, cans, building materials).

5. The MSW characterized（描述）in the EPA report includes waste from residential,

commercial, institutional, and industrial sources.

V. Reading comprehension.

Hazardous Materials

A Primer of Hazardous Materials

A hazardous material is one that poses some form of danger to humans or the environment. More specifically, according to RCRA, materials are considered hazardous if they cause or significantly contribute to an increase in mortality or an increase in serious irreversible, or incapacitating reversible illness; or pose a substantial present or potential hazard to human health or the environment when improperly treated, stored, transported, disposed of, or otherwise managed.

Classifications for Hazardous Materials

Many hazardous materials may fall into more than one category. Descriptions of the hazards posed by these materials are classified into seven basic types:

1. Flammable/Combustible（易燃物）—ignites easily and burns rapidly.

2. Explosive（易爆物）/Reactive—explosive chemicals produce a sudden, almost instantaneous（瞬间的）release of pressure, gas, and heat when subjected to abrupt shock, high temperature, or an ignition source; reactive chemicals vigorously undergo a chemical change under conditions of shock, pressure, or temperature.

3. Sensitizer（致敏物质）—on first exposure causes little or no reaction in humans or test animals; but on repeated exposure may cause a marked response not necessarily limited to the contact site. Skin sensitization is the most common form; respiratory sensitization to a few chemicals also occurs.

4. Corrosive（侵蚀性物质）—causes visible destruction of or irreversible alterations in living tissue by chemical action at the site of contact.

5. Irritant（刺激物）—noncorrosive chemicals that causes a reversible inflammatory effect on living tissue by chemical action at the site of contact as a function of concentration or duration of exposure.

6. Carcinogen（致癌物质）—either causes cancer in humans, or, because it causes cancer in animals, is considered capable of causing cancer in humans.

7. Toxic（有毒物质）—poisonous to living organisms when they are ingested, inhaled, or absorbed through the skin.

Decide whether each of the following statements is true (T) or false (F) according to the material.

( ) 1. Descriptions of the hazards posed by these materials are classified into six basic types.

( ) 2. Explosive chemicals may produce a sudden, almost instantaneous release of pressure, gas, and heat.

( ) 3. Irritant refers to corrosive chemicals that cause a reversible inflammatory effect on living tissue.

( ) 4. Carcinogen only can cause cancer in animals.

( ) 5. Toxic materials are poisonous to living organisms when they are ingested, inhaled, or absorbed through the skin.

Unit 2　Summary of Solid Waste Management

Text　Integrated Waste Management

Integrated Waste Management

An integrated waste management system combines two or more of the following processes:

source reduction
reuse
recycling
composting
incineration
emissions
landfill burial or encasement

Source Reduction（源头减量）

Probably the most important component of any effective integrated waste management system is a strategy for reducing the amount of refuse entering the waste stream in the first place. Source reduction includes any action that reduces the volume（体积）or toxicity of solid waste prior to recycling or disposal.

Reuse（再利用）

It is easy to fall into the pattern of believing that newer is always better. Unfortunately, many perfectly good products are routed into the waste stream every day to make way for new products. Even some heavily used items might still have value—they just need to be repaired or refurbished（翻新）.

Recycling（资源化）

Each person in the United States generates, on average, 4~5 pounds of MSW per day. A large part of this waste can undergo resource recovery, where the materials are salvaged as raw materials through recycling or composting or as energy through incineration. Among the most easily recycled materials are paper, metals, glass, and plastics. Recyclable materials can be recovered from mixed solid waste by employing hand-sorting and various types of machinery—air classifiers, magnets, cyclones, trommels, crushers, grinders, and balers.

Composting（堆肥）

Composting is the controlled biological process of turning organic waste into a soil conditioner. Composting produces a nutrient-rich soil additive called compost, which is used to improve soil quality (and thus plant growth) by increasing nutrient availability, water-holding capacity, aeration, and biological activity. Besides its value as a method of waste reduction and soil treatment, composting has recently been discovered to have important applications in pollution prevention and control. Compost applied to creek, lake, or river embankments or on roadsides and hillsides reduces silting and erosion. It can also reduce heavy metals and organic contaminants in storm water runoff, preventing contamination of water. Compost degrades or completely eliminates such contaminants as hydrocarbons and pesticides. In addition, the use of mature compost can suppress plant diseases.

Incineration（焚化）

In addition to recycling and composting, some resources can be recovered from the MSW stream through incineration. One of the purposes of incineration is to increase the useful life of available landfills and minimize odor and sanitation problems. An efficient mass-burn incinerator can reduce the solid waste going into a landfill by as much as 80% ～ 90% in volume and 65% ～ 75% in mass.

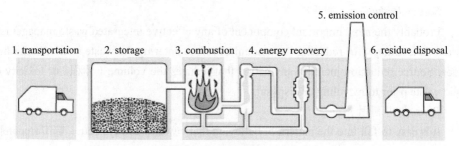

Figure 1　The process of incineration

The process of incineration in Figure 1 composes of transportation, storage, combustion, energy recovery, emission control and disposal of residue.

1. Transportation—MSW is collected and delivered to the mass-burn facility.

2. Storage—Waste is transferred to a storage pit or tipping floor.

3. Combustion—A conveyor or crane transfers the waste to the hopper, which feeds the waste into the furnace. Secondary combustion chambers aid a more complete combustion.

4. Energy recovery—The heat from combustion is transferred to water in pipes, which turns into steam. Steam is used directly for processes or to generate electricity.

5. Emission control—Dry and wet scrubbers and other air-pollution-control devices, such as electrostatic precipitators and fabric filters, remove some of the acid gases and particulates from the exhaust.

6. Disposal of residue—The ash from burning and the residue from scrubbers and other pollution-control devices are disposed of in a landfill.

Emissions（排放）

Most energy-recovery facilities use complicated combustion-control systems designed to optimize combustion, minimize ash for disposal, and optimize clean burning by reducing the formation of products of incomplete combustion (PICs). Some of the waste that goes through the incineration process, however, might exit the system in one of the following forms: combustion gases（气体燃烧）, particulate emissions（颗粒物排放）, fly ash（飞灰）and bottom ash（底灰）.

Landfill Burial or Encasement（填埋）

Other waste management processes cannot completely eliminate the need for landfills, no matter how successfully these processes are implemented. Landfills will always be necessary if the MSW stream continues to have the same constituents as it currently does.

Two types of modern landfills exist: sanitary and secure.

Sanitary Landfills（卫生填埋场）

Sanitary landfills have filters and liners to prevent contamination of soil; leachate collection and monitoring systems to prevent the contamination of groundwater; and methods of collecting methane gas to prevent explosive pockets from building up in the landfill（See Figure 2）.

Sanitary landfills can be relatively benign to the environment. Some landfills collect the methane gas from decomposition and use it as a fuel source for the neighboring communities.

Secure Landfills（安全填埋场）

Secure landfills are authorized to accept toxic waste and have much stricter safety precautions（措施）than sanitary landfills. Different types of toxic waste are entombed in separate chambers, and a careful inventory is kept of what items are buried there. Rather than having just a compacted-soil base and a permeable geomembrane liner (like that of sanitary landfills), secure landfills have two compacted base layers—clay and soil—and an impermeable plastic liner protecting the landfill walls. The cap is also deeper and stronger, and a groundwater monitoring well is installed to help detect leaks (See Figure 3).

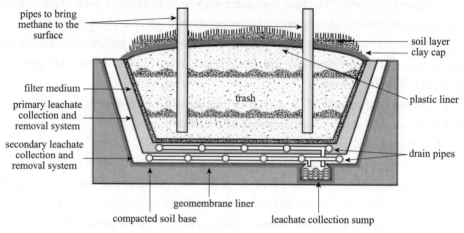

Figure 2 Cross-section of a sanitary landfill

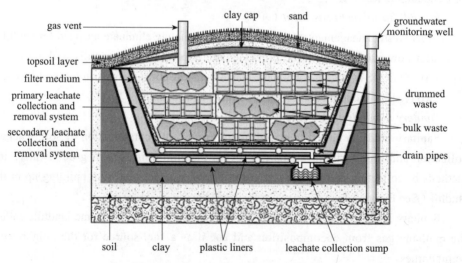

Figure 3 Cross-section of a secure landfill

Secure landfills require the highest degree of care in their design and upkeep(维护). They are designed for the indefinite storage of toxic chemicals, and great vigilance(警觉) must be taken to prevent leaks. Because of the nature and concentration of the chemicals buried there, a secure landfill that leaks can cause deadly contamination of the surrounding environment, much more serious than contamination from a sanitary or unsanitary landfill.

Words and Expressions

  integrated ['intigretid] *adj.* 综合的；互相协调的
  salvage ['sælvidʒ] *vt.* 抢救；捞回
  recover [ri'kʌvə] *vt.* 重新获得；恢复；还原
  liner ['lainə] *n.* 防渗层；衬板；衬垫；衬套
  leachate ['litʃet] *n.* 浸出液；沥出物；沥出液；渗漏污水
  detect [di'tekt] *vt.* 检测
  impermeable [im'pɜ:miəb(ə)l] *adj.* 防渗的；不渗透性的；（对水等）不能渗透的
  secure [si'kjʊə; si'kjɔ:] *adj.* 安全的

  integrated solid waste management 固体废弃物综合治理
  landfill burial or encasement 垃圾填埋场填埋或封装
  make way for 为……让路
  heavily used 大量使用的
  resource recovery 资源回收
  soil conditioner 土壤结构改良剂
  source reduction 源头减量
  geomembrane liner 隔泥网膜
  impermeable plastic liner 防渗塑料膜衬套
  monitoring well 监测井

Notes

1. Probably the most important component of any effective integrated waste management system is a strategy for reducing the amount of refuse entering the waste stream in the first place. 在源头就减少进入废物流的垃圾量的措施也许是任何有效的废物综合治理系统中最重要的组成部分。
   - in the first place 一开始；起初
   - component 组成部分
   - strategy 措施；策略
2. A large part of this waste can undergo resource recovery, where the materials are salvaged as raw materials through recycling or composting or as energy through incineration. 这种

垃圾大部分都可以进行资源回收，即让这些材料通过回收和堆肥被重新作为可利用的原材料或通过焚化来提供能源。
- resource recovery= recycling 资源化；资源回收；循环利用
- undergo 经受
- be salvaged 被挽救

3. Recyclable materials can be recovered from mixed solid waste by employing hand-sorting and various types of machinery—air classifiers, magnets, cyclones, trommels, crushers, grinders, and balers. 人们可以通过手工分选和应用风力分级器、磁铁、旋风分离器、矿石筛、破碎机、研磨机和打包机等各种类型的机械装置对混合固体废弃物进行处理以重新获得可回收利用的材料。
- hand-sorting 手工分选
- air classifiers 风力分级器
- magnet 磁铁
- cyclone 旋风分离器
- trammel 矿石筛
- crusher 破碎机；碎渣机；碎石机
- grinder 研磨机
- baler 打包机

4. Composting produces a nutrient-rich soil additive called compost, which is used to improve soil quality (and thus plant growth) by increasing nutrient availability, water-holding capacity, aeration, and biological activity. 堆肥产生一种营养丰富的土壤改良剂，此改良剂也被称为混合肥料，它可通过增加土壤的养分有效性、蓄水能力、曝气程度和生物活性来改善土质（所以植物才会生长）。
- 句中 which 引导非限制性定语从句。介词 by 短语作方式状语
- additive 添加物；辅助物
- nutrient availability 养分有效性
- water-holding capacity 蓄水能力

5. Compost applied to creek, lake, or river embankments or on roadsides and hillsides reduces silting and erosion. It can also reduce heavy metals and organic contaminants in storm water runoff, preventing contamination of water. 在小溪、湖泊、河堤、路边和山坡上实施堆肥可以减少土壤的淤积和流失，还能减少雨水径流中的重金属和有机污染物，防止水污染。
- silting 淤积；淤塞
- erosion 侵蚀

6. One of the purposes of incineration is to increase the useful life of available landfills and

minimize odor and sanitation problems. 焚化的目的之一是延长已有的垃圾填埋场的使用寿命，并把气味浓度和污染程度减到最低。
- minimize 使减到最少

  eg. At every stage we try to minimize waste. 在每个阶段，我们都努力减少浪费。

  We had about ten hours' warning, so we were able to minimize the effects of the flood. 由于在十小时前发出了警报，我们能够把水灾的破坏减至最低程度。

7. Rather than having just a compacted-soil base and a permeable geomembrane liner (like that of sanitary landfills), secure landfills have two compacted base layers—clay and soil—and an impermeable plastic liner protecting the landfill walls. 安全填埋场并非只有一个压实土基底层和可渗水的隔泥网膜（像卫生填埋场那样），而是有两个压实土基底层——黏土层和褐黄色土层——和一层不透水的塑料防渗层来保护垃圾填埋坑的坑壁。
- rather than 而不是
- Rather than having just a compacted-soil base and a permeable geomembrane
- liner (like that of sanitary landfills) 分词短语作主句的状语。
- permeable 透水的；可渗透的
- impermeable 不透水的；防渗的
- clay 黏土
- soil 地面土壤；褐黄色土壤

Exercises

I. Decide whether each of the following statements is true (T) or false (F) according to the material.

( ) 1. An integrated waste management system contains only one step.

( ) 2. Plastics is easy to be recycled.

( ) 3. Composting is the uncontrolled biological process of turning organic waste into a soil conditioner.

( ) 4. Secure landfills have one compacted base layers.

( ) 5. Leaks from a secure landfill can cause more deadly contamination of the surrounding environment than from a sanitary or unsanitary landfill.

II. Translate the following words or phrases into English.

1. 综合治理 _____   2. 源头减量 _____

3. 再利用；重新使用 _____   4. 堆肥 _____

5. 排放控制 _____   6. 旋风分离器 _____

7. 焚化炉 _____   8. 飞灰 _____

9. 织物过滤器 _____   10. 安全填埋场 _____

III. Match the words and expressions with their meanings.

( ) 1. PICs   a. 蓄水能力
( ) 2. sanitary landfill   b. 运移；搬运
( ) 3. water-holding capacity   c. 发电
( ) 4. generate electricity   d. 安全填埋场
( ) 5. transport   e. 未完全燃烧产品
( ) 6. sanitary landfills   f. 研磨机；磨床
( ) 7. crusher   g. 碳氢化合物
( ) 8. hydrocarbon   h. 使最优化；完善
( ) 9. optimize   i. 碎渣机
( ) 10. grinder   j. 卫生填埋场

IV. Match the words and expressions with their meanings.

( ) 1. resource recovery   a. reject
( ) 2. unfortunately   b. burning
( ) 3. refuse   c. recycling
( ) 4. combustion   d. safe
( ) 5. secure   e. luckily

V. Translate the following sentences into Chinese.

1. Source reduction includes any action that reduces the volume（体积）or toxicity of solid waste prior to recycling or disposal.

2. Among the most easily recycled materials are paper, metals, glass, and plastics.

3. The process of incineration in Figure 1 composes of transportation, storage, combustion, energy recovery, emission control and disposal of residue.

4. Some landfills collect the methane gas from decomposition and use it as a fuel source for the neighboring communities.

5. The cap of secure landfill is also deeper and stronger, and a groundwater monitoring well is installed to help detect leaks.

Part IV　Abstract Writing

科技论文摘要写作

科技论文摘要是科技论文的一个重要组成部分。它是以简要的方式，清晰地概括论文的主体内容。摘要的作用在于方便读者在短时间内了解论文的要点和梗概，并为读者是否决定深入研读全文提供参考信息。

国内科技论文摘要一般都有汉英两个部分。汉语摘要用于国内学术同行间的交流，而英文摘要则是便于国际间的学术交流和合作。英文摘要是利于学术论文的国际检索和提高论文引用率的一个必备条件。

高职理工科学生在科技论文写作方面起码应具备三项技能：一是能读懂本专业及相关专业的科技论文摘要。二是了解科技论文摘要的基本体例。三是能够按规则写出符合基本要求的英文摘要。此三项技能相辅相成，对其中一项的熟练掌握都有利于其他两项的提高。

一、类型

科技论文摘要的类型一般分为陈述性的（Descriptive）和资料性的（Informational）两类。陈述性摘要只说明论文、书籍或文章的主题，多半不介绍内容。资料性的摘要除了介绍主题外，还应介绍文章的要点和各个要点的主要内容。它可以包括三个组成部分：①点明主题，解析文章或书籍的目的或意图；②介绍主要内容，使读者迅速了解文章或书籍的概貌；③提出结论或建议，以供读者参考。

二、写作指要

科技论文摘要写作的总体原则是简要清晰、术语规范。英汉两种文本要信息对等，语言地道连贯。具体而言，英文科技论文摘要是一种很规范的文体，它有自身的规范体例。从总体层面上看，它有形式和内容两个方面的要求。从句子层面上看，它又有一些较为鲜明的文法特征以及常用的句型。

（一）形式要求

1. 摘要的题名

英文题名以短语为主要形式，尤以名词短语（noun phrase）最常见。其中有四点要注意：一是注意词序，要先厘定中心词，再进行前后修饰；二是字数不应过长，一般不超过两行为宜；三是冠词，凡可用可不用的地方均可不用；最后一个是题目中字母的大小写，通行格式为首字母大写，但虚词如冠词、连词以及介词全部小写。

2. 作者与单位的英译

中国人名按汉语拼音拼写；其他非英语国家人名按作者自己提供的罗马字母拼法拼写。单位名称要写全（由小到大），并附地址和邮政编码，确保联系方便。

3. 摘要字数

形式要求主要是指字数要求。学术期刊英文科技论文摘要允许的字数范围在 100～500 字。一般来讲，250 字较为适宜。

违反此要求出现的常见错误主要有摘要太短或太长两种。例如摘要字数只有数十个字，或摘要远远超出摘要字数的上限。这些形式上的错误主要原因是违背了内容要求，因为内容决定形式。

（二）内容要求

（1）摘要需包含作者研究的目的、方法、简要分析和结果。

（2）摘要不得重复论文内的具体数据以及引用其他资料所提供的数据。

（3）读者能够通过摘要而在不通读全文的情况下就可以了解全文要点和梗概。

上述内容3方面的基本要求从根本上框定了形式上的构架。摘要太短是因为仅仅扣住内容要求中的第一点，而忽略了内容要求的第3点，未能提供充分的必要信息。而摘要太长则一般是因为违反了内容要求的第2点。

（三）文法特征

1. 摘要的时态

英文摘要时态的运用也以简练为佳，常用一般现在时、一般过去时，少用现在完成时、过去完成时，进行时态和其他复合时态基本不用。一般现在时用于说明研究目的、叙述研究内容、描述结果、得出结论、提出建议或讨论等。涉及公认事实、自然规律、永恒真理等，当然要用一般现在时。

2. 摘要的语态

采用何种语态，既要考虑摘要的特点，又要满足表达的需要。一篇摘要很短，尽量不要随便混用，更不要在一个句子里混用。需要指出的是，为了体现行文的客观性，在科技论文摘要中谓语动词用被动语态的很常见，当然现在主张摘要中谓语动词尽量采用主动语态的也越来越多，因其有助于文字清晰、简洁及表达有力。

3. 摘要的人称

原来摘要的首句多用第三人称"this paper..."等开头，现在倾向于采用更简洁的被动语态或原形动词开头，行文时最好不用第一人称。

（四）常用句式

1. 表目的的句型

—The purpose of the paper is…

—The primary goal of the research is…

—The author attempted the set of experiments with a view to demonstrating certain phenomena…

2. 用于分析的句型

—The experiment consists of three steps, which are described in…

—This is a working theory based on the idea that…

—The method used in this study is known as…

3. 用于结论的句型

—In conclusion, we stated that…

—The results of the experiment indicate that…

—The pioneer studies we have attempted seem to indicate that…

三、实例分析

Example 1:

Analysis of Chemical Constituents in Waste Gas Pollutant of Paint Industry by Gas Chromatography-mass Spectrometry and Chemometric Resolution Method

Abstract: Objective: The chemical constituents in waste gas pollutant of paint industry were analyzed. **Method:** Analysis of Chemical constituents in waste gas pollutant of paint industry was performed with two-dimensional gas chromatography-mass spectrometry (GC/MS) data coupled with chemometric resolution method (CRM). **Result:** By means of two-dimensional data, 49 chemical components in waste gas pollutant of paint industry were parsed from 25 chromatographic peaks and 40 compounds were determined qualitatively and quantitatively, accounting for 92.32% total contents of waste gas pollutant of paint industry. **Conclusion:** The experimental results show that the main components in waste gas pollutant of paint industry were benzene series and alkane, accounting for 46.79% and 27.23%, respectively. The chemometric resolution method decreased the requirement of chromatographic resolution and provided the new way of analyzing complex unknown system quickly and precisely.

Key words: paint industry; waste gas; chemometric resolution method; gas chromatography-mass spectrometry

解析：该摘要用 5 个句子，共计 138 个英文单字，概括了油漆行业废气化学成分的 GC-MS 测定与化学计量解析。为便于快速阅读，作者用黑体标示出目的、方法、实验结果和结论，全文简单明了。

摘要题目为"油漆行业废气化学成分的 GC-MS 测定与化学计量解析"。题目的英文表述的形式为"analysis of…by…"，其中 analysis 后的 3 个介词短语"of chemical constituents""in waste gas pollutant""of paint industry"按从后向前的顺序，依次修饰限定前面的所指，简单讲，从 analysis 后的 of 开始，直到 industry 都是 analysis 的限定语，对该词的内涵进行界定。

"全文目的在于分析油漆行业废气污染物化学成分。"摘要中用一句话概述，并且在陈述时采用的是被动语态"The chemical constituents…were analyzed"。英文中被动语态常用于科技文献，它可以保持行文的客观性。

"方法采用 GC/MS 法，来分离测定油漆行业废气污染物的化学成分，利用化学计量学解析法（CRM）对重叠的色谱峰进行解析，得到各成分的纯色谱曲线和质谱，通过质谱库对解析的纯组分进行定性，用解析色谱曲线积分法进行定量。"方法描述同样一句话，也是采用被动语态"Analysis of Chemical constituents…was performed with…"效果同上。

"实验结果是从 25 个色谱峰中解析出了 49 个组分，鉴定了其中 40 个化合物，占总含量的 92.32%。"表述实验结果用了一个句子，并用具有鲜明文体特征的表示方法的短语，例如"By means of these methods…"以及"…accounting for…"。句子同样采用被动语态，例如"…were parsed…"以及"…were determined qualitatively and quantitatively…"，保持行文的客观语气。

结论是："废气污染物的主要成分为苯系物和烷烃，分别占总含量的 46.79% 和 27.23%。"该法降低了对色谱分离度的要求，为快速、准确解析复杂未知体系提供了新的途径。结论表述为两个句子，采用文体特征鲜明的表结论性的句式"The experimental results show that…"以及"…provided the new way of…"。

关键词有 4 个，油漆行业、废气、化学计量学解析法和气相色谱-质谱，较为准确地反映了论文的要点。

Example 2:

Analysis of PAHs in Atmosphere by Gas Chromatography-mass Spectrometry and Chemometric Resolution Method

Abstract: Analysis of PAHs in atmosphere was performed with two-dimensional gas chromatography-mass spectrometry (GC/MS) data coupled with chemometric resolution method (CRM).By means of these methods two-dimensional data, 14 chemical components of PAHs in atmosphere were parsed from 21 chromatographic peaks. The chemometric resolution method decreased the requirement of chromatographic resolution and provided the new way of analyzing complex unknown system quickly and precisely.

Key words: PAHs; chemometric resolution method; gas chromatography-mass spectrometry

解析：该摘要包含3个句子，共计62个英文单字。全文陈述模式是"以……方法，得到……结果，此方法的意义是……"全文简练。

摘要题目为"GC-MS测定与化学计量学解析大气中的多环芳烃"。题目的英文表述仍采取"analysis of…by…"的形式，同理可知analysis后的of短语是该词的限定修饰语。需要指出的是，题目内出现了"PAHs"这个专业名词的缩写，专业名词的含义对专业学生来说并非难事，但要注意行业内术语的规范和一致性。

"实验方法采用GC/MS法测定大气中多环芳烃，利用化学计量学解析法（CRM）对色谱峰进行解析，得到各成分的纯色谱曲线和质谱，通过质谱库对解析的纯组分进行定性，用解析色谱曲线积分法进行定量。从21个色谱峰中解析出了14种多环芳烃。"用于方法的描述句采用了文体色彩显著的表达式"Analysis of…was performed with…"以及"By means of these methods…"。

"该法降低了对色谱柱分离度的要求，为快速、准确解析复杂未知体系提供了新的途径。"本句表明了研究方法的意义，采用的句式"The chemometric resolution method decrease… and provided the new way of analyzing…"。

关键词有3个，多环芳烃、化学计量学解析法和气相色谱-质谱，集中体现了全文的重要信息。

Exercises

I. Put the sentences into right order and find out the key words of the abstract you have formed.

(A)
Environmental Management and Waste Disposal in Colleges and Universities

① The treatment of various wastes from daily work, learning and life for the teachers and students, as well as the campus environment planning and management, is an important part of running modern universities.

② This paper discussed the environment issue, the particularity of environment management and some problems in treatment of wastes in colleges and universities.

③ As a special community of society, colleges and universities take the responsibility of personnel training, scientific research, social service and cultural inheritance and innovation.

The right order: _____.
The key words: _____.

(B)
The Super-ministers Reform of Ecological Environment Management in Western Countries and the Enlightenments for Our Country

① This paper mainly analyzes the practices of the ecological environment management in British, American, Japan, Canada and other western countries to provide beneficial enlightenments to the super-ministries reform of ecological environment management in China.

② The super-ministries reform of environmental protection in accordance with the principles of ecological system has been gradually carvied out by the main developed countries' environmental protection system since the 1980s.

③ The establishment and reform of the ecological environment management system in western countries is based on the developmental trend and complexity of the ecological environment problems.

The right order: _____.
The key words: _____.

II. Fill in the blanks as follows.

(A)
Mechanism for Transformation of Environment Criteria into Environmental Standards in China

Abstract: In order to provide scientific basis for developing more reasonable and effective environmental standards (ESs), China has taken its own environmental criteria (EC) research strategy since recent years. For the purpose of developing a transformation mechanism from EC to ESs suitable for China, current EC and ESs in China were compared with those in some developed countries, such as the USA, Canada and Australia. Four common characteristics of transformation from EC to ESs in developed countries and corresponding problems in China were drawn through the comparison, based on which a frame work for transformation from EC to ESs was established. The framework was composed by three parts: Relevant national and local agencies and responsibilities; steps for transformation including affecting factors and their interactions; policies for solving current problems in China.

The key words: _____.
The aim of the article is: _____.

(B)
Analysis of Chemical Constituents in Waste Gas Pollutant of Coatings Industry by Chemometric Resolution Method

Abstract: The chemical constituents in waste gas pollutant of Coatings industry were complex unknown system, which is not applied to qualitative and quantitative analysis. Analysis of Chemical constituents in waste gas pollutant of coatings industry was performed with two-dimensional gas chromatography-mass spectrometry (GC/MS) data coupled with chemometric resolution method (CRM).By these methods of two-dimensional data, 47 chemical components in waste gas pollutant of coatings industry were parsed from 20 Chromatographic Peaks and 39 compounds were determined qualitatively and quantitatively, accounting for 91.02% of total contents of waste gas pollutant of coatings industry, respectively. The experimental results show that the main components in waste gas pollutant of coatings industry were benzene series, accounting for 47.89%. The chemometric resolution method decreased the requirement of chromatographic resolution and provided the new way of analyzing complex unknown system quickly and precisely.

The key words: _____.
The aim of the article is: _____.

III. Translate the titles into Chinese.

1. Environmental Management and Waste Disposal in Colleges and Universities

2. The Super-ministers Reform of Ecological Environment Management in Western Countries and the Enlightenments for our Country

3. Mechanism for Transformation of Environment Criteria into Environmental Standards in China

4. Analysis of Chemical Constituents in Waste Gas Pollutant of Coatings Industry by Chemometric Resolution Method

IV. Translate the abstract into Chinese.

The Application of Environment Impact Assessment in HSE Management System

Abstract: HSE (Health, Safety and Environment) management system has become the general management system implemented in international petroleum and petrochemical industry. Petrochina progressively conducts the implementation of HSE management system and continuously makes exploration and practice. This article starts from interrelation of environmental impact assessment and HSE management system in oil companies, conducts feasibility analysis of employing environmental impact assessment in HSE management system, and draws the following conclusion: to improve the efficiency and effectiveness of business management and promote long-term development of the enterprises, we should fully integrate environmental management with environmental impact assessment.

Key words: HSE management system; system construction; environmental impact assessment; environment management

V. Translation.

生物流化床在废水处理中的应用进展

摘要：本文介绍了生物流态化技术在废水处理方面的应用历史与发展概况，以及近年来出现的一些新型生物流化床反应器的操作原理和结构特点。指出生物流化床应向着降低能耗、适应不同水质的处理方向发展。

关键词：废水；流化床（fluidization bed）；生物处理

附 录

附1　环境工程专业用语参考

（一）环境学总论

原生环境 primary environment
次生环境 secondary environment
生态示范区 ecological demonstrate area
环境地质学 environmental geology
环境地球化学 environmental geo-chemistry
环境土壤学 environmental soil science
环境微生物学 environmental microbiology
环境危机 environmental crisis
环境保护 environmental protection
环境预测 environmental forecasting
环境自净 environmental self-purification
环境效应 environmental effect
环境容量 environmental capacity
环境演化 evolution of environment
环境舒适度 environmental comfort
环境背景值（本底值）environmental backgroud value
环境保护产业（环保产业）environmental production industry
环境壁垒（绿色壁垒）environmental barrier
绿色革命 green revolution
可持续发展 sustainable development
第三类环境问题（社会环境问题）the third environmental problem
悬浮物 suspended solids
可更新资源 renewable resources
不可更新资源 non-renewable resources
自然保护区 natural reserve area
防护林 protection forest
公害 public nuisance
矿山公害 mining nuisance
工业废水 industrial waste water
矿山废水 mining drainage
生活饮用水 domestic potable water
草原退化 grassland degeneration
沙漠化 desertification
人口压力 population pressure
人口净增率 rate of population
全球环境监测系统 global environment monitoring system, GEMS
中国环境保护工作方针 Chinese policy for environment protection
"三同时"原则 principle of "the three at the same time"
二噁英公害 dioxine nuisance

马斯河谷烟雾事件 disaster in Meuse Valley
多诺拉烟雾事件 disaster in Donora
伦敦烟雾事件 disaster in London
水俣病事件 minamata disease incident
骨痛病事件 itai-itai disaster incident
一次污染物 primary pollutant
二次污染物 secondary pollutant
全球性污染 global pollution
排污收费 pollution charge
洛杉矶光化学烟雾事件 Los Angeles photochemical smog episode
四日市哮喘事件 Yokkaichi asthma episode
米糠油事件 Yusho disease incident

（二）环境工程学

环境污染综合防治 integrated prevention and control of pollution
环境功能区划 environmental function zoning
稀释比 dilution ratio
迁移 transfer
紊流扩散 turbulent diffusion
氧亏（亏氧量）oxygen deficit
复氧 reaeration
溶解氧下垂曲线 dissolved-oxygen sag curve
饱和溶解氧 saturated dissolved
无污染燃料 pollution-free fuel
燃烧 combustion
空气—燃料比 air-to-fuel ratio
烟气分析 analysis of flue gas
煤的综合利用 comprehensive utilization of coal
脱硫 desulfurization
除尘效率 particle collection efficiency
分割粒径 cut diameter for particles
压力损失（压力降）pressure drop
机械除尘器 mechanical collector
重力沉降室 gravity settling chamber
惯性除尘器 inertial dust separator
旋风除尘器 cyclone collector
气布比 air-to-cloth ratio
机械振动清灰袋式除尘器 bag house with shake cleaning
逆气流清灰袋式除尘器 bag house with reverse-flow cleaning
脉冲喷吹清灰袋式除尘器 bag house with pulse-jet cleaning
静电除尘 electrostatic precipitator (ESP)
高压脉冲静电除尘器 pulse charging electrostatic precipitator
湿式静电除尘器 wet electrostatic precipitator
重力喷雾洗涤器 gravitational spray scrubber
旋风洗涤器 centrifugal scrubber
中心喷雾旋风洗涤器 cyclone spray scrubber
泡沫洗涤塔 foam tower scrubber
填料床洗涤器 packed bed scrubber
文丘里洗涤器 venturi scrubber
双膜理论 two-film theory
气膜控制 gas film control
液膜控制 liquid film control

穿透曲线 break through curve
催化剂 catalyst
催化剂中毒 poisoning of catalyst
烟气脱硫 flue gas desulfurization (FGD)
湿法脱硫 wet process of FGD
石灰—石灰石法脱硫 desulfurization by lime and limestone
氨吸收法脱硫 ammonia process of FGD
干法脱硫 dry process FGD
吸收法控制氮氧化物 control of NO_x by absorption
水吸收法脱氮 control of NO_x by absorption process with water
酸吸收法脱氮 control of NO_x by absorption process with acid
碱吸收法脱氮 control of NO_x by absorption process with alkali
回流式旋风除尘器 reverse-flow cyclone collector
直流旋风除尘器 straight-through cyclone collector
多管旋风除尘器 multiple cyclone collector
过滤除尘器 filter
袋式除尘器 bag house
滤料 filtration media
电晕放电 corona discharge
驱进速度 drift velocity
集尘极 collecting electrode
板间距 distance between collecting electrodes
电极清灰 removal of collected particle from electrodes
宽间距静电除尘器 wide space electrostatic precipitator
双区静电除尘器（两段式电除尘器） two-stage electrostatic precipitator

湿式除尘器 wet collector of particulates
吸附法控制氮氧化物 control of NO_x by adsorption
分子筛吸附法脱氮 control of NO_x by adsorption process with molecular sieve
硅胶吸附法脱氮 control of NO_x by adsorption process with silica gel
气体生物净化 biotreatment of gaseous pollutant
生物过滤器 biofilter
汽车尾气污染 pollution of automobile exhaust gal
生物脱臭 biotreatment of oder
集气罩 capture hood
烟囱有效排放高度 effective height of emission
清洁生产 cleaner production
矿山废水 mining drainage
电镀废水 electroplating wastewater
给水处理厂 water treatment plant
污水处理厂 wastewater treatment
给水（污水）处理构筑物 water (sewage) treatment structure
污水集水井 swage joining well
废水调节池 wastewater flow equalization basin
格栅 grill
筛网 grid screen
沉砂池 grit settling tank
立式圆形沉砂池 vertical circular grit settling tank
圆形周边运动沉砂池 circular perimeter flow grit settling tank
重力排砂 grit discharge by gravity
水力提升排砂 grit discharge with hydraulic

elevator
水力旋流器 hydraulic cyclone
沉淀池 settling tank
重力沉淀池 gravity settling tank
竖流折板絮凝池 vertical table flap flocculating tank
机械搅拌絮凝池 mechanical mixing flocculating tank
加压溶气气浮法 pressure dissolved-air floatation
过滤池 filter
重力过滤法 gravity filtration process
压力过滤法 pressure filtration process
真空过滤法 vacuum filtration process
快滤池 rapid filtration
慢滤池 slow filtration
接触滤池 contact filter
双向滤池 bidirectional filter
双层滤料滤池 double layer filter
无阀滤池 non-valve filter
虹吸滤池 siphon filter
压力滤池 pressure filter
V形滤池 aquazur V-filter
砂滤 sand filtration
微滤机 microstrainer
滤池冲洗强度 backwashing intensity of filter
滤层 filter material layer
滤料承托层 holding layer for filter material
斜板隔油沉淀池 oil trap with slope plank
冷却塔 cooling tower
湿式氧化法 wet oxidation process
反应池 reaction basin
叶轮搅拌器 turbine mixer
膜分离法 membrane separation method
剩余污泥 surplus sludge

初次沉淀池 primary sedimentation basin
完全混合曝气池 completely mixed aeration basin
曝气沉砂池 aeration grit settling tank
平流式沉砂池 horizontal grit settling tank
浓缩沉淀池 thickening settling tank
斜板（斜管）沉淀池 sloping plank (pipe) settling tank
辐流式沉淀池 radial settling tank
平流式沉淀池 horizontal settling tank
竖流式沉淀池 vertical settling tank
悬浮污泥澄清池 suspended sludge clarifier
脉冲澄清池 pulse clarifier
水力循环澄清池 hydraulic circulating clarifier
颗粒自由沉降 particle free sediment
絮凝沉降 flocculation sedimentation
拥挤沉降 hindered sedimentation
气浮池 floatation basin
微电解法 micro electroanalysis
半渗透膜 semi-permeable membrane
电渗析 electrodialysis
反渗透 reverse osmosis
离子交换膜 ion exchange membrane
萃取 extraction
汽提 stripping
吹脱法 blow-off method
臭氧氧化法 ozonation
臭氧发生器 ozonator
磁分离法 magnetic isolation method
光催化氧化 optical catalysis oxidation
软化水处理 softening water treatment
石灰—纯碱软化法 lime-sodium carbonate softening method
废水好氧/厌氧处理 biological aerobic/an-

aerobic treatment of wastewater
微生物内源代谢 microorganism intrinsic metabolism
微生物合成代谢 microorganism synthetic metabolism
基质分解代谢 substrate degradation metabolism
活性污泥法 activated sludge process
回流污泥 return sludge
曝气池 aeration basin/tank/pond
推流式曝气池 plug-flow aeration basin
二次沉淀池 secondary sedimentation basin
污泥沉降比 sludge settling ratio
污泥容积指数 sludge velum index
污泥负荷 volume loading
纯氧曝气法 oxygen aeration method
扩散曝气设备 diffusion aerator
机械曝气装置 mechanical aerator
曝气时间 aeration time
污泥龄 sludge age
活性污泥培养 activated sludge culture
活性污泥驯化 domestication of activated sludge
粉末炭活性污泥法 powdered carbon activated sludge process
污泥膨胀 sludge bulking
生物滤池 biological filter
高负荷生物滤池 high-loading biological filter
水力负荷 hydraulic loading
有机负荷 organic loading
塔式生物滤池 tower biological filer
生物转盘 biological rotating disc
生物流化床 biological fluidized bed
活性生物滤池 activated biofilter

化粪池 septic tank
污水硝化脱氮处理 nitrogen removal from wastewater by nitrification
污水反硝化脱氮处理 nitrogen removal from wastewater by denitrification
污水硝化—反硝化脱氮处理 nitrogen removal from wastewater by nitridenitrification
土地处理系统 land treatment system
氧化塘 oxidation pond
好氧塘 aerobic pond
兼性塘 facultative pond
厌氧塘 anaerobic pond
曝气氧化塘 aerated oxidation pond
间歇循环延时曝气活性污泥法（ICEAS）intermittent cyclic extended aeration system
需氧池—间歇池联合工艺（DAT-IAT）demand aeration tank intermittent aeration tank system
A/O 工艺 anoxic/ oxic
分段曝气法 step aeration method
延时曝气法 extended aeration method
加速曝气法 accelerant aeration method
深井曝气法 deep well aeration method
鼓风曝气装置 blast aerator
射流曝气设备 efflux aerator
表面曝气装置 surface aerator
A^2/O 工艺 phostriop process
Phostriop 工艺 Phostriop process
Bardenpho 工艺 Bardenpho process
Phoredox 工艺 Phoredox process
UCT 工艺 university of cape town
VIP 工艺 Virginia initiative plant

厌氧生物滤池 (AF) anaerobic filter
厌氧接触法 anaerobic contact process
厌氧生物转盘 anaerobic biological rotating disc
两相厌氧消化 two-phase anaerobic digest
序批式间歇反应器 series batch reactor
氧化沟 oxidation ditch
上流式厌氧污泥床 upflow anaerobic sludge blanket
MSBR modified sequencing batch reactor
消毒 disinfection
灭菌 sterilization
加氯机 chlorinator
氯化消毒 chlorization disinfection
漂白粉消毒 disinfection by bleaching powder
紫外线消毒 disinfection with ultraviolet rays
加氯消毒 disinfection by chlorine
液氯 liquified chlorine gas
需氯量 chlorine demand
余氯 chlorine residual
游离性余氯 free chlorine residual
化合性余氯 combined chlorine residual
折点加氯 chlorination breakpoint
过氧化氢消毒 disinfection by hydrogen peroxide
除味 taste removal
除臭 odor removal
脱色 decoloration
生污泥 undigested sludge
熟污泥 digested sludge
污泥处置 disposal of sludge
污泥综合利用 comprehensive utilization of sludge
真空过滤法 vacuum flotation process
污泥浓缩 sludge thickening
污泥消化 sludge digestion

污泥脱水 sludge dewatering
污泥干化 sludge drying
污泥焚烧 sludge incineration
真空过滤机脱水 dewatering by vacuum filter
板框压滤机脱水 dewatering by plate frame press filter
辊轧式脱水机脱水 dewatering by roll press
带式压滤机脱水 dewatering by belt press filter
离心式脱水机脱水 dewatering by centrifuge
中温消化处理 middle temperature digestive treatment
高温消化处理 high temperature digestive treatment
污泥堆肥发酵处理 sludge composting and fermentation
污泥浓缩池 sludge thickener
污泥消化池 sludge digestion tank
污泥产气率 gas production rate of sludge
污泥干化场 sludge drying bed
固体废物 solid wastes
城市生活垃圾 municipal solid wastes
城市生活垃圾堆放处置法 dumping of municipal solid wastes
城市生活垃圾卫生填埋法 sanitary landfilling of municipal solid wastes
城市生活垃圾焚烧法 incineration of municipal solid wastes
城市生活垃圾分类 sorting of municipal solid wastes
城市生活垃圾收集 collection of municipal solid wastes
垃圾收费 refuse taxing
废电池 used battery
有毒有害工业固体废物 toxic industrial

wastes
医疗废物 health care wastes
堆肥 composting
填埋场 landfill
渗滤液 leachate treatment
焚烧炉 incineration furnaces
助燃空气系统 air injection system
余热利用 heat utilization
焚烧灰渣 ash
水泥固化技术 cement solidification
石灰固化 lime solidification
沥青固化技术 asphalt solidification

固体废物预处理 preliminary treatment of solid wastes
破碎 crushing of solid wastes
筛分 screening of solid wastes
风力分选 wind separation
放射性固体废物 radioactive solid waste
声级计 sound level meter
消声室 anechoic room; anechoic chamber; dead room
混响室 reverberation room
隔声 sound insulation
吸声 muffler

（三）环境地学

水圈 hydrosphere
水循环 water circulation
地面水（地表水）surface water
水位 water level
下渗（入渗）sinking
蒸发 evaporation
最高水位 highest water level
冲刷 washout
最低水位 lowest water level
平均水位 average water level
警戒水位 warning water level
流速 flow velocity
流量 discharge
洪水期 flood season
枯水期 low-water season
含水层 aquifer
隔水层（不透水层）aquiclude
透水层 permeable stratum
层间水 interlayer water
承压水（有压层间水）confined water

自流水 artesian water
孔隙水 void water
岩溶水（喀斯特水）karst water
径流 runoff flow
地表径流 runoff
地下水 ground water
流域保护 water basin protection
淡水 fresh water
咸水 saltwater
降水；沉淀 precipitation
降水量 amount of precipitation
降水强度 intensity of precipitation
水环境容量 carrying capacity of water environment
水土流失（土壤侵蚀）soil and water loss
点源污染 point source pollution
面源污染 non-point source pollution
扩散 diffusion
涡流 eddy current
涡流扩散 eddy diffusion

富营养化废水 eutrophic waste-water
污水 sewage
漫灌 flood irritation
水底沉积物（底质或底泥）benthal deposit
总固体 total solids
悬浮固体 suspended solids
总溶解固体 total dissolved solids
河流复氧常数 constant of river reoxygenation
湖泊酸化 lake acidification
富营养化 eutrophication
富营养湖 eutrophic lake
中营养湖 mesotrophic lake
贫营养湖 oligotrophic lake
水库 reservoir
海洋处置 sea disposal
海底采样 sea floor sample
赤潮（红潮）red tide
海水淡化 desalination of seawater
海底沉积物 sea bottom sediment
海洋倾倒 ocean dumping
水质 water quality
水资源综合利用 water resource integrated utilization
水土保持 soil and water conservation
河道整治 channel improvement
水污染毒性生物评价 biological assessment of water pollution toxicity
水利工程 hydro-engineering
水体自净 self-purification of water body
水环境保护功能区（水质功能区）function-al district of water environment
土地处理系统 land treatment system
土地沙漠化 land desertification
土壤肥力 soil fertility
土壤酸碱度 soil acidity and alkalinity

土壤污染防治 prevention and treatment of soil pollution
土壤盐渍化（土壤盐碱化）soil salination
土壤酸化 soil acidification
母质（土壤母质或成土母质）parent material
土壤剖面 soil profile
腐殖质化 humification
淋溶作用 leaching
土壤改良 soil improvement
土壤粒级 soil separate
土壤质地 soil texture
缓冲作用 buffering/buffer action
缓冲剂 buffering agent/buffer
缓冲容量 buffer capacity
盐基饱和度 base saturation percentage
灌溉 irrigation
富里酸 fuvic acid
胡敏素 humin
土壤团聚体 soil aggregate
土壤退化（土壤贫瘠化）soil degeneration
土壤地带性 soil zonality
污水灌溉 wastewater irrigation
臭氧层 ozone layer
降水 precipitation
降水量 rainfall
降水强度 precipitation intensity
大气环境容量 atmospheric environmental capacity
事后评价 afterwards assessment
烟尘消除 elimination of smoke and dust
温室效应 greenhouse effect
大气扩散 atmospheric diffusion
烟羽（烟流或羽流）plume
逆温 inversion

（四）环境化学

甲基汞 methyl mercury
镉米 cadmium rice
农药残留 pesticide residue
有机氯农药 organochlorine pesticide
有机磷农药 organophosphorous pesticide
氨基甲酸酯杀虫剂 carbamate insecticide
拟除虫菊酯杀虫剂 pyrethroid insecticide
植物生长调节剂 growth regulator
化学致癌物 chemical carcinogen
表面活性剂 surfactant
多氯联苯类 polychlorinated biphenyls，PCBs
多环芳烃类 polyaromtic hydrocarbon，PAH
催化（催化作用）catalysis
臭氧化 ozonization
光化学氧化剂 photochemical oxidant
过氧乙酰硝酸酯 peroxyacetyl nitrate，PAN
干沉降 dry deposition
湿沉降 wet deposition
光化学烟雾 photochemical smog
大气光化学 atmospheric photochemistry
降水化学 precipitation chemistry
气溶胶化学 aerosol chemistry
悬浮颗粒物 suspended particulate
总悬浮颗粒物 total suspended particulates（TSP）
飘尘（可吸入颗粒物或可吸入尘）airborne particle
降尘（落尘）dustfall；falling dust
气溶胶 aerosol
水质 water quality
盐度 salinity
氧化还原电位 oxidation-reduction potential；redox potential
溶解氧 dissolved oxygen
化学需氧量 chemical oxygen demand
生化需氧量 biochemical oxygen demand
总有机碳 total organic carbon
溶解度 solubility
聚集 aggregation
絮凝 flocculation
凝聚 coagulation
离子交换 ion exchange
萃取 extraction
缓冲溶液 buffer solution
氧平衡模式（氧垂曲线）oxygen balance model
吸收剂（吸附剂）absorbent
活性炭 active carbon
氧化剂 oxidant
还原剂 reductant
胶团 micelle
胶体溶液 colloidal solution
脱硫剂 desulfurization agent
电渗析 electrodialysis
萃取剂 extracting agent
过滤 filter
絮凝剂 flocculant；flocculating agent
无机絮凝剂 inorganic flocculant
有机高分子絮凝剂 organic polymer floc-culant
中和法 neutralization
反渗透膜 reverse osmosis membrane
硅胶 silica gel
蒸汽蒸馏 steam distillation

超滤膜 ultrafilter membrane
灵敏度 sensitivity
准确度 accuracy
精密度 precision
可靠性 reliability
检测限 detection limit
相对误差 relative error
绝对误差 absolute error
偶然误差 accidental error
平均偏差 mean deviation
采样误差 sampling error
标准溶液 standard solution
标准物质 standard substance
允许误差 allowable error
允许浓度 allowable concentration
微量分析 microanalysis
痕量分析 trace analysis
现场分析 in-situ analysis
仪器分析 instrumental analysis
水质分析 water quality analysis
比色分析 colorimetric analysis
沉降分析 sedimentation analysis
自动分析 automatic analysis
原子吸收分光光度法 atomic absorption spectrophotometry
原子吸收分光光度计 atomic absorption spectrophotometer
原子荧光光谱法 atomic fluorescence spectrometry
原子荧光光谱仪 atomic fluorescence spec-trometer
电化学分析法 electrochemical method
高效液相色谱法 high performance liquid chromatography
高效液相色谱仪 high performance liquid chromatograph
气相色谱分析 gas chromatography
气相色谱仪 gas chromatograph
采样器 sampler
大气采样器 air sampler
底泥采样器 sediment sampler
pH 计 pH meter
湿度计 hygrometer
固定大气污染源 stationary sources of air pollution
移动大气污染源 mobile sources of air pollution
固定式水污染源 stationary sources of water pollution
移动式水污染源 mobile sources of water pollution
污染负荷 pollution load
污染源调查 survey of pollution sources
无污染工艺 pollution-free technology
无污染装置 pollution-free installation
污染物总量控制 total amount control of pollution
水质参数 water quality parameter
水温 water temperature
色度 color index
透明度 transparency
混浊度 turbidity
硬度 hardness
感官污染指标 sensuous pollution index
毒理学污染指标 physical pollution index
化学污染指标 chemical pollution index
细菌学污染指标 bacteriological pollution index
毒理学污染指标 toxicological pollution index
城市污水 municipal sewage
生活污水 domestic sewage
工业废水 industrial wastewater

常规分析指标 index of routine analysis
环境监测 environmental monitoring
过程监测 course monitoring
污染物排放标准 pollution discharge standard
总量排放标准 total amount of pollution discharge standard
优先监测 priority monitoring
环境优先污染物 environmental priority pollutant
总固体 total solids
可吸入微粒（可吸入尘和飘尘） inhale particles
浊度计 turbidimeter
实验室质量控制 laboratory quality control
空白实验值 blank value
平行样 duplicate samples
再现性（重现性） reproducibility
重复性 repeatability
回收率 recovery rate
检出限 detection limit
冷原子吸收法 cold-vapor atomic absorption method
紫外吸收光谱法 ultraviolet absorption spectrophotometry
重量分析 gravimetric analysis

内标法 internal marker method
定性分析 qualitative analysis
定量分析 quantitive analysis
试样前处理 pre-treatment
均值 mean value
标准差 standard error
方差 variation
回归分析 regression analysis
相关分析 correlation analysis
相关系数 correlation coefficient
系统误差 systematic error
随机误差 random error
有效数字 valid figure
农药残留分析 pesticide residue analysis
排污收费 effluent charge
室内空气污染 indoor air pollution
水体自净 self-purification of water body
水土保持 soil and water conservation
水土流失 soil erosion
土壤修复 soil-remediation
生物修复 bioremediation
光降解 photodegradation
温室气体 greenhouse gases
总量收费 total quantity charge
超临界流体 supercritical fluid
土壤采样 soil pollution

（五）环境物理学

光辐射（光）visible radiation
红外线 infrared ray
紫外线 ultraviolet ray
灭菌灯 bactericidal lamp
光污染 light pollution

噪声污染 noise pollution
混响 reverberation
听力损失 hearing loss
绝对湿度 absolute humidity
相对湿度 relative humidity

饱和度 saturation ratio
冷凝 condensation
露点温度 dew point temperature

热辐射 thermal radiation
比热 specific heat
空气调节 air conditioning

（六）环境法学

通风 ventilation

环境法学 science of environmental law
环境保护法 environmental protection law
公害法 public nuisance law
环境行政法规 administrative regulations of environment
环境部门规章 departmental rules of environment
污染物排放标准 pollutant discharge stand-ard
"三同时"制度 three simultaneity system
排污审报登记制度 declaration and registration system of pollution ischarge
排污许可证制度 permit system of pollutant discharge
排污收费制度 system of effluent
限期治理制度 system of eliminating and controlling environmental pollution within a prescribed time
现场检查制度 system of on-site inspection
环境污染事故报告制度 system of environmental pollution accident reporting
《中华人民共和国环境保护法》Environmental Protection Law of the People's Republic of China
《中华人民共和国水污染防治法》law of the People's Republic of China on prevention and control of water pollution
《中华人民共和国大气污染防治法》law of the People's Republic of China on prevention and control of atmospheric pollution
《中华人民共和国环境噪声污染防治法》law of the People's Republic of China on prevention and control of pollution from environmental noise
《中华人民共和国固体废物污染环境法》law of the People's Republic of China on prevention and control of environmental pollution by solid waste
《中华人民共和国海洋环境保护法》marine environment protection law of the People's Republic of China
《全国生态环境建设规划》national eco-environmental construction plan
《全国生态环境保护纲要》national compendium on eco-environmental protection
地表水环境质量标准 environmental quality standard for surface water
地下水质量标准 quality standard for ground water
农业灌溉水质标准 standard for irrigation water quality
污水综合排放标准 integrated wastewater discharge standard
大气污染物综合排放标准 integrated emission standard of air pollutants

附2 参考译文与参考答案

Part I 水处理

Unit 1 水处理概述

Text 水污染物质及其来源

排入地表水的污染物质可以分成几个大的类别。

点源

生活污水和工业废水都被称为水污染点源,这是因为它们通常都会被一个由各种管道或渠沟形成的网络收集起来,并集中到某个排放点排入收纳水体。生活污水包含有家庭住所、学校、办公楼和商店里人们在日常生活中产生的废水。城市污水这一术语是指含有工业污水的生活污水。总之,点源污染是可以通过在排入自然水体之前进行适当的处理而减少或者去除的。

非点源(面源)

城市径流和农田径流以来自于多处排放点为特征。我们称它们为非点源(面源)。通常这类污水会在地表横流或者沿着天然的排水沟流入距离最近的水体。即便当城市径流和农田径流被汇集到管道中,通常也会在尽可能短的运输距离内排放出来,如此一来,要在每个排放口进行水处理在经济上不合算。很多面源污染发生在暴雨或者春季融雪期间,这时极大的水流量使得此时的面源污染的处理更要困难得多。要减少面源污染通常需要变更土地的利用方式和提高教育水平。

耗氧物质

所有在收纳水体中能够通过消耗水中的溶解氧而被氧化的物质都可以叫做耗氧物质。这种物质通常是指能够进行生物降解的有机物质,但也包括某些无机化合物。这些耗氧物质对水中溶解氧(缩写为 DO)的消耗对必须依赖氧而存活的更高形式的水生生命的生存构成了威胁。溶解氧浓度的临界水平在不同物种中是不一样的,比如溪点红蛙需要在每升水约含 7.5mg 溶解氧的水体中生存，而鲤鱼则要在每升水中约含 3mg 溶解氧的水体中才能存活。通常最受青睐的商业鱼类和猎用鱼类的生存都需要较高的溶解氧浓度。生活污水中的耗氧物质主要来源于人类排泄物和餐饮垃圾。在很多产生耗氧废物的产业中值得注意的是食品加工业和造纸业。几乎所有的自然产生的从非点源排入水体的有机物质，比如动物的排泄物、作物残茬和树叶等，都促成了溶解氧的消耗。

营养物质

氮和磷这两种最主要的营养物质在浓度过高时也被看成污染物质。一切生命要生存都离不开它们，江河湖海中的食物链中必需要有它们的存在。但当水中的营养物质过多，食物链也会受到极大干扰，这会导致某些生物体以牺牲其他生物体为代价而激增，就会出现问题。过多的营养物质会导致水中的藻类大量生长，而反过来当它们死亡并沉到水底后又变成了耗氧物质。营养物质即营养盐的主要来源是以磷为基本成分的洗涤剂、化肥和食品加工过程中产生的废弃物。

病原微生物（致病菌）

废水中的微生物包括由动物或病人排泄和分泌出来的细菌、病毒和原生动物。一旦他们被排入地表水，就会使得水体不再适合饮用。如果病菌的浓度足够高，人们在这样的水中游泳和捕鱼也不安全了。某些甲壳类动物能将致病菌聚集在其身体组织内，使其毒性程度远大于其所处水体的毒性程度，因此它们就会变得有毒。

悬浮物

被废水携带进入某个收纳水体的有机颗粒和无机颗粒就叫做悬浮物或悬浮固体（缩写为 SS）。当水因流入池塘或湖泊而流速减缓时，这些颗粒很多就会下沉到水底变成沉积物。在通常的用法中，沉积物这个词汇的含义还包括被水冲刷侵蚀而形成的土壤颗粒，哪怕这些土壤颗粒还没有沉到水底。人们发现在很多地表水体中不易沉淀的胶原颗粒会导致水体混浊。有机悬浮颗粒会产生需氧量。无机悬浮颗粒是从某些工厂企业排放出来的，但多半是水土流失造成的，这在那些伐木、露天采矿和建筑施工活动区域情况尤其严重。过多的沉积物在湖泊和水库中堆积，降低了这些湖泊和水库的水利调节功能。甚至在流速很快的山涧溪流中，因采矿业和伐木作业而产生的沉积物也会破坏很多水生生物的生存环境（生态栖息地）。例如，鲑鱼卵只能在松散的砾石河床里发育和孵化。而鹅卵石之间的气孔里塞满了沉积物，导致鱼卵窒息而死，鲑鱼的数量就随之减少了。

有毒金属和有毒的有机化合物

农田径流通常含有施加给农作物的除草剂。城市径流是很多水体中的铅和锌的主要来源。铅源于使用含铅汽油的汽车的尾气排放，而锌源于轮胎磨损。很多工业废水既含有有毒金属，也含有有毒的有机物质。如果大量排放，会使得收纳水体在很长时间内丧失用途。由于沿岸工厂产生的剧毒和持久性有机物的大量排放,弗吉尼亚州的詹姆士河下游现已仅可用作航运通道了。许多有毒化合物在食物链中聚集,使得人们再也不能安全地食用这些水里的鱼类和甲壳类动物了。因此,哪怕只含有少量的这类有毒物质的水体也不能与自然生态系统相容了,在很多方面也不适合人类使用了。

热

尽管人们并不经常把热看成一种污染物质,但是那些供电行业的人们还是非常清楚废热排放导致的问题。而且,很多工业生产使用过后排出的水要比收纳水体的水热得多。在一些环境里,水温的增加是有益的。比如,在某些地方,给水增暖可以促进蛤蚌和牡蛎生长。从另一方面看,水温的升高也可能会有负面的影响。许多像鲑鱼和鳟鱼这样重要的商业鱼类和猎用鱼类只能在冷水中生存。在某些情况下,水电站废热的排放有可能彻底阻住鳟鱼的洄游。另外,较高的水温也提高了存在耗氧废弃物的水域的耗氧率。

Key to the exercises

I. Choose the best choices to fill in the blanks:
1. D 2. C 3. C 4. B 5. D

II. Decide whether each of the following statements is true (T) or false (F) according to the text.
1～5: T T F F T

III. Translate the following words or phrases into English.

1. pollutant
2. treat, treatment; dispose, disposal
3. discharge
4. sewage; waste water
5. microorganism
6. nutrient
7. algae
8. particle
9. waste heat
10. biodegradable

IV. Match the words and expressions with their meanings.
1～5: j e i g c   6～10: h b f d a

V. Match the words and expressions with their meanings.
1～6: c a b e f d

VI. Translate the following sentences into Chinese.

1. 要减少面源污染通常需要变更土地的利用方式和提高教育水平。
2. 营养盐的主要来源是以磷为基本成分的洗涤剂、化肥和食品加工过程中产生的废弃物。
3. 某些甲壳类动物能将致病菌聚集在其身体组织内，使其毒性程度远大于其所处水体的毒性程度，所以就会变得有毒。
4. 甚至在流速很高的山涧溪流中，因采矿业和伐木作业而产生的沉积物也已经破坏了很多水生生物的生存环境（生态栖息地）。
5. 很多工业废水既含有有毒金属，也含有有毒的有机物质。

Part I 水处理

Unit 2 一级处理

Text A 污水预处理（1）

排入地表水的范围广泛的污染物可以被分为几大等级。

随着文明的进步与城市的发展，水质日益恶化。生活污水和工业废水不断流入排水沟和下水道，并最终流入附近的河道。对主要城市来说，这些污水排放通常足以破坏大量的水体。

污水特性

生活污水、工业废水和渗透物排放到污水沟管系统，其中渗透物本身并不需要处理，它仅增加了污水总量，甚至在一定程度上能稀释污水。这些污水的处理，随工业大小、类型及污水量的不同而有所不同。

随着时间的推移，不同社区的生活污水的水质和水量有明显的变化。一个小社区的日变化情况（见图1和表1）显示了典型的生活污水日变化系数。

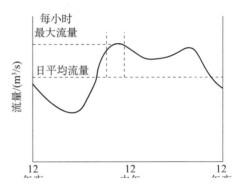

图 1 小社区日流量变化

表 1 典型生活污水的特性

参数	生活污水典型值
生化需氧量/（mg/L）	250
悬浮物/（mg/L）	220
磷/（mg/L）	8
有机氮和氨氮/（mg/L）	40
pH 值	6.8
化学需氧量/（mg/L）	500
总固体颗粒物浓度/（mg/L）	270

为了达到排放标准，污水处理系统一般分为三级：

一级处理：应用物理处理法去除污水中较大的固体颗粒，并且均和废水。

二级处理：应用生物处理法去除大量的生化需氧物质。

三级处理：应用物理处理法、生物处理法和化学处理法等，去除污水中的磷等营养物及有机污染物，除臭、脱色，并进行进一步氧化。

格栅与沉砂

未经处理的污水排放到河道，其中悬浮物最不受欢迎。于是，格栅是社区处理污水的第一种方式，如今也是污水厂处理污水的第一个步骤。典型的格栅（见图2略）由一组钢条组成，间隙约2.5cm。现代污水处理厂的格栅用于去除污水中较大的悬浮物，这些物质可能会损坏设备或阻碍进一步的处理。一些老的污水处理厂使用人工清理格栅，但几乎所有新的污水处理厂都使用机械格栅。当格栅被充分堵塞，导致钢条前的水位提升时，清理耙就会被激活。

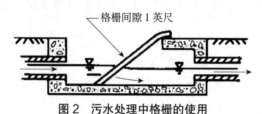

图2 污水处理中格栅的使用

很多污水处理厂的第二个处理步骤是用破碎机，即圆磨床来研磨固体，这些固体在穿过格栅后，变成直径约0.3 cm或更小的碎片。典型的破碎机见图3。

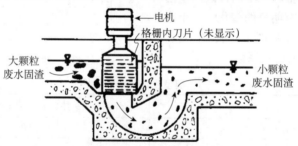

图3 用来研磨大颗粒固体的破碎机

第三个处理步骤是去除污水中的沙砾。沙砾会损坏如水泵和流量计等设备，因此必须去除。最常见的沉砂池是一种宽型渠道，这种池子的水流速度慢到足够使相对密度大的沙砾沉淀。沙砾的密度是大多数有机颗粒的2.5倍，因此沉淀得更快。沉砂池的功能是从污水中分离相对密度较大的无机颗粒，如沙砾等，而有机颗粒将会在之后进行处理。分离出来的沙砾可用作填料，无须进行额外处理。

Key to the exercises

I. Answer the following questions after reading the text.

1. Because this discharge was often enough to destroy even a large body of water.
2. The objective of primary treatment is to remove nonhomogenizable solids and homogenize the remaining effluent.
3. According to the Figure1, we know the maximum sewage flow is almost at noon, the minimum sewage flow is at about 6:00 in the morning.
4. Because it can remove materials that might damage equipment or hinder further treatment.
5. The objective of a grit chamber is to remove sand and grit without removing organic materials.

II. Decide whether each of the following statements is true (T) or false (F) according to the text.

1～5: F F F T T

III. Translate the following words or phrases into English.

1. sewer   2. Primary treatment   3. cleaning rake   4. pump   5. raw sewage
6. screen   7. ammonia   8. floating material   9. comminutor   10. COD

IV. Translate the following sentences into Chinese.

1. 排入污水沟管系统的污水包括生活污水、工业废水以及渗透物。
2. 一级处理：应用物理处理法去除污水中较大的固体颗粒，并均和废水。
3. 于是，格栅是社区处理污水的第一种方式，如今也是污水厂处理污水的第一个步骤。
4. 很多污水处理厂的第二个处理步骤是用破碎机，即圆磨床来研磨固体，这些固体在穿过格栅后，变成直径约 0.3cm 或更小的碎片。
5. 最常见的沉砂池是一种宽型渠道，这种池子的水流速度慢到足够使相对密度大的沙砾沉淀。

V. Match the words and expressions with their meanings.

1～5: f d b a i   6～10: c h j e g

VI. Best choices.

1. D   2. A   3. A   4. B   5. C

VII. Fill in the following blanks.

图略。

Motor（发动机）

大颗粒废水固渣（Wastewater with large solids）

Wastewater with small solids（小颗粒废水固渣）

Part I 水处理

Unit 2 一级处理

Text B 污水预处理（2）

沉淀池

大部分污水处理厂在修建了沉砂池后还会接着修建沉淀池，用来尽可能多地沉淀固体颗粒。于是，水力停留时间要长，湍流要保持在最低限度。停留时间为污水平均流入池中所花费的总时间，它以填满沉淀池所需的时间来计算。例如，池容量为 100 m³，流速为 2 m³/min，停留时间为 100/2=50 min。一般来说，沉淀池分为两种：矩形沉淀池（见图 1）和圆形沉淀池（见图 2）。

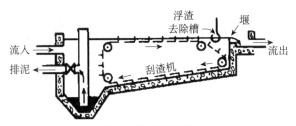

图 1 矩形沉淀池

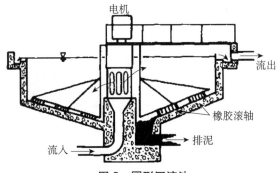

图 2 圆形沉淀池

水中的固体物质沉淀到池底后通过排泥管排出池外，同时澄清的液体溢出三角堰。三角堰能使液体一直围绕水池平均排放。沉淀池又称为澄清池。沉淀池与格栅、沉砂池一起被称为初级澄清池。沉入初级澄清池并被排出的固体被称为初次沉淀污泥。

初次沉淀污泥通常容易腐化发臭，含水率高，不易处理。它需先稳定，减少进一步分解和脱水以便于处理。其他的固体颗粒也必须同样处理后才能排放。

气浮

气浮是从液体中去除油脂和悬浮颗粒的一种方法，是处理含油脂悬浮物的最有效的方法之一。其中，溶气气浮应用最为广泛。首先，污水在一个密封的罐内与空气进行加压混合。然后，通过泄压阀将污水送入气浮池（见图3）。由于突然减压，形成了直径只有 50～100 μm 的细微气泡，悬浮颗粒和乳化油就黏附于气泡周围而随其上浮。通常的做法是使用化学物质来提高浮选性能。与混凝剂一样，浮选剂应是无害的，回收的固体物质经常用作动物饲料配方。

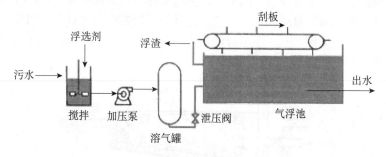

图3 溶气气浮法

另一种方法是取一部分除油后出水（10%～30%）回流（见图4）。所有的气浮池都有一种机械装置来去除沉淀于池底的固体，这种装置通常是螺旋形的传送带，一般置于锥底。与沉淀池相比，溶气气浮的主要优点是：细小悬浮物可以更快被去除。

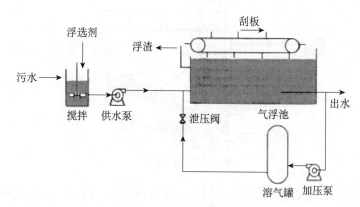

图4 回流溶气气浮法

据报道，溶气气浮法的性能取决于几个因素，其中最重要的因素是固体颗粒的浓度。含固量越高，去除率通常也更强。其他影响其性能效率的因素是气固比，即减压后空气释放量与原水中悬浮固体量的比值。最佳气固比通常是由实验室规模测试所决定。

溶气气浮运作成功的关键因素包括：适当的pH值（通常在4.5～6，最常见的降低蛋白质溶解度和拆分乳液的pH值为5）、适当的流速及训练有素的操作员不间断在场。

Key to the exercises

I. Answer the following questions after reading the text.

1. Retention time is the total time an average slug of water spends in the tank and is calculated as the time required to fill the tank.
2. Because it distributes the liquid discharge equally all the way around a tank.
3. The waste stream is first pressurized with air in a closed tank. After passing through a pressure-reduction valve, the wastewater enters the flotation tank where, due to the sudden reduction in pressure, minute air bubbles in the order of 50-100 microns in diameter are formed. As the bubbles rise to the surface, the suspended solids and oil or grease particles adhere to them and are carried upwards.
4. The most important that DAF systems performance depend on is the solids concentration.
5. The optimum A/S data is determined by bench scale tests.

II. Decide whether each of the following statements is true (T) or false (F) according to the text.

1～5: T F T F T

III. Translate the following words or phrases into English.

1. primary clarifier   2. sludge   3. V-notch weir   4. flotation
5. pressure pump   6. formulation   7. coagulant
8. concentration   9. sedimentation 10. removal efficiencies

IV. Translate the following sentences into Chinese.

1. 停留时间为污水平均流入池中所花费的总时间，它以填满沉淀池所需的时间来计算。
2. 固体沉淀到池底后通过排泥管排出池外，同时澄清的液体溢出三角堰。三角堰能使液体一直围绕水池平均排放。
3. 悬浮颗粒和乳化油就黏附于气泡周围而随其上浮。
4. 所有的气浮池都有一种机械装置来去除沉淀于池底的固体，这种装置通常是螺旋形的传送带，一般置于锥底。
5. 据报道，溶气气浮法的性能取决于几个因素，其中最重要的因素是固体颗粒的浓度。

V. Match the words and expressions with their meanings.
1～5: c j b a i   6～10: d h e f g

VI. Best choices.
1. D   2. D   3. A   4. B   5. D

VII. Fill in the following blanks.
图略。

流入（Influent）　　　　　　　　（Sludge）排泥
浮渣去除槽（Scum trough）　　　（Weir）堰
（刮渣机）Sludge scrapers　　　　Effluent（流出）

浮选剂（Flotation aids）　　　　　Floating sludge（浮渣）
（Pressure pump）加压泵　　　　Retention tank（溶气罐）
（Pressure reduction value）泄压阀　（Flotation tank）气浮池

Part I 水处理

Unit 3 二级水处理

Text A 好氧废水处理——活性污泥系统

由初级澄清池排出的水已经去除了大部分的有机物，但仍含有一些高分子有机物，在微生物的作用下，高分子有机物分解会产生生化需氧量。必须减少对氧的需求以降低有机污染物浓度（氧量消耗），否则在收纳水体排放时可能会出现不可接受的状况（污染物浓度超标）。废水初级/一级处理目的是去除悬浮固体物质和沙粒，而二级处理旨在降低 BOD 值。

如图所示，一个活性污泥系统，包括来自初级澄清池的废水和大量微生物。进入曝气池的气泡为需氧生物的生存提供必要的氧气。微生物接触废水中的溶解性有机物，吸附它们，并最终将这些有机物分解为二氧化碳、水、一些稳定的化合物以及新生繁殖的微生物。新生微生物的产生速度相对较慢，并且会占用大部分的曝气池容积。

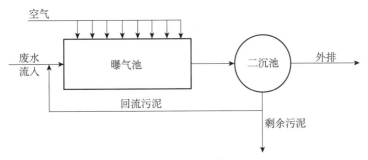

图 1 活性污泥系统流程

当作为微生物食物的大部分有机物被消耗殆尽时，微生物便在沉淀池与液体分离（沉淀池有时也被称为二级或最终澄清池）。沉淀池里的微生物没有食物来源，自身活性被饥饿激活，最终变成活性污泥。上清液经由出水堰，可能排入收纳水体。遗留下的微生物（现可称为回流活性污泥）由泵送回曝气池前端，在曝气池前端有由初级澄清池注入曝气池的污水，微生物们可以在污水中的有机化合物里寻觅到更多的食物，此过程一遍又一遍循环反复。随着持续的污泥泵送和净水排放，活性污泥处理连续不间断地处理污水。

活性污泥处理法会产生比实际需要量更多的微生物，如果这些微生物没有外排，会使得其浓度增加从而导致固体物质堵塞系统，因此必须排除这些微生物。清理这些剩余活性污泥是污水处理中最困难的环节之一。

活性污泥系统的设计基于有机物的数量和荷载量，以及食物中可用的微生物量。食物-微生物（F/M）比例是一个主要的设计参数，F和M很难准确测得，但是通过流入曝气池中的BOD和SS（悬浮固体）能够分别估算出它们的大概值。进行曝气的污水和微生物组合称为混合液，在曝气池中的SS则就是混合液悬浮固体浓度。进水BOD和MLSS的比例，即F/M比例，是整个系统的荷载量，是按照每天每单位质量（磅或公斤）MLSS所对应的BOD质量（磅或公斤）来计算的。

相对较小的F/M比例，或少量食物对应大量微生物，以及长时间的曝气周期（曝气池中的停留时间）都可引起高负荷处理，其原因在于微生物能够最大限度地利用可用食物，具有这些特点的系统称为延时曝气系统。延时曝气系统可广泛应用于孤立的废水来源，如一些小型新建住宅区或度假酒店。延时曝气系统几乎不会产生多余的生物量和需要进行处理的剩余活性污泥。

废水的二级处理通常包括一个生物净化过程，如活性污泥，二级处理去除很大一部分的BOD和剩余的固体。典型废水水质如表1所示：

表1 典型废水水质

	原水水质	一级处理后出水水质	二级处理后出水水质
BOD/(mg/L)	250	175	15
SS/(mg/L)	220	60	15
P/(mg/L)	8	7	6

二级处理后的废水符合国内目前BOD和SS排污标准，但是磷含量仍然超标。如要去除包括无机磷和含氮化合物等无机化合物，则需要高级或三级废水处理。

Key to the exercises

I. Answer the following questions after reading the text.

1. The objective of secondary treatment is to remove BOD, whereas the objective of primary treatment is to remove solids.
2. Air bubbled into this aeration tank provides the necessary oxygen for survival of the aerobic organisms.
3. When most of the organic material, which is food for the microorganisms, has been used up, the microorganisms are separated from the liquid in a settling tank, sometimes called a secondary or final clarifier. The microorganisms remaining in the

settling tank have no food available, become hungry, and are thus activated; hence the term activated sludge.
4. Activated sludge treatment produces more microorganisms than necessary and if the microorganisms are not removed, their concentration will soon increase and clog the system with solids.
5. Both F and M are difficult to measure accurately, but may be approximated by influent BOD and SS in the aeration tank, respectively.

II. Decide whether each of the following statements is true (T) or false (F) according to the text.
1～5: F F T F F

III. Translate the following words or phrases into English.
1. primary clarifier
2. primary treatment
3. activated sludge
4. aerobic organism
5. secondary / final clarifier
6. receiving water
7. return activated sludge
8. extended aeration system
9. secondary treatment
10. effluent standard

IV. Translate the following sentences into Chinese.
1. 废水初级/一级处理目的是去除悬浮固体物质和沙粒，而二级处理旨在降低 BOD 值。
2. 进入曝气池的气泡为需氧生物的生存提供必要的氧气。
3. 沉淀池里的微生物没有食物来源，自身活性被饥饿激活，最终变成活性污泥。
4. 活性污泥处理法会产生比实际需要量更多的微生物，如果这些微生物没有外排，会使得其浓度增加从而导致固体物堵塞系统。
5. 进行曝气的污水和微生物组合称为混合液，在曝气池中的 SS 则就是混合液悬浮固体浓度（MLSS）。

V. Match the words and expressions with their meanings.
1～5: f d g b i   6～10: a c j e h

VI. Best choices.
1. D   2. C   3. D   4. D   5. A

VII. Fill in the following blanks.
图略。
Waste influent（废水流入） （Aeration tank）曝气池
（Final Clarffier）二沉池 Return activated sludge（回流污泥）
Waste activated sludge（剩余污泥）

Part I 水处理

Unit 3 二级水处理

Text B 厌氧废水处理

厌氧废水处理是利用生物制剂在无氧环境中去除废水中的污染物。经过这样处理，废水就能安全地排放到外界环境中去。厌氧废水处理过程中使用的生物制剂为微生物，这些微生物能消耗或分解污泥中的生物降解材料以及污水过滤后的废水中的固体部分。

因为微生物在处理过程中所起的作用，厌氧废水处理又称为厌氧消化，从根本上来说，处理方式实际上是"消化"水中的污染物。作为减少污水和剩余食物中有机物的一个极佳方法，厌氧消化是生物废水处理系统中具有代表性的一个组成部分。

厌氧处理系统包含三个复杂的过程，过程中除了产生其它生成物之外，生物消化还会产生沼气。首先，水解脂肪、纤维素和蛋白质。由栖息细菌产生的细胞外酶将大分子水解产物分解成更小和更易消化的分子。接下来，脂肪酸被进一步分解，一些兼性厌氧菌进行这一分解过程。最后，产甲烷菌消化吸收这些脂肪酸，从而形成沼气。

厌氧处理通常在放置于地面或地下的密封容器中进行。在最初的污泥分解阶段，主要由细菌构成的微生物将废料转变为有机酸、氨气、氢气和二氧化碳。在厌氧废水处理的最后阶段，一种被称为产甲烷菌的单细胞微生物将污泥中的残余物转化成含有沼气和二氧化碳的生物气体。

厌氧废水处理的一个额外效益是减少气体排放。处理过程中产生的生物气体可作为替代能源用于做饭、照明、取暖和发动机燃料。换句话说，收集和利用厌氧消化过程中产生的沼气和二氧化碳，这些生物气体就不会被排放至大气中。

科学家们普遍认为地球大气层中高浓度的沼气和二氧化碳（又名温室气体）是导致全球变暖的元凶。这个被称为温室效应的理论认为，因为这些气体将来自太阳的热能留在了大气层，所以导致全球气温升高。虽然这一理论引发了不少争议，但是利用生物气体替代化石燃料的做法具有一定实用价值。

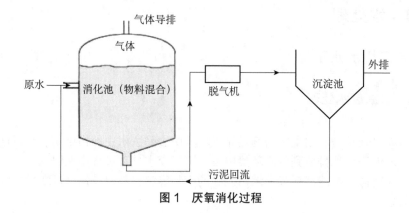

图 1 厌氧消化过程

Key to the exercises

I. Answer the following questions after reading the text.
1. The biological agents used in the process are microorganisms that consume or break down biodegradable materials in sludge, or the solid portion of wastewater following its filtration from polluted water.
2. Anaerobic wastewater treatment is also known as anaerobic digestion due to the action of the microorganisms. That is, they are essentially "digesting" the polluted parts of the water.
3. The first stage is the hydrolysis of lipids, cellulose, and protein. Extracellular enzymes produced by the inhabiting bacteria break down these macromolecules into smaller and more digestible forms. Next, these molecules are decomposed into fatty acids. This decomposition is performed by several facultative and anaerobic bacteria. Finally, methanogenic bacteria digest these fatty acids, resulting in the formation of methane gas.
4. Usually, the anaerobic process takes place in sealed tanks, located either above or below the ground.
5. Anaerobic wastewater treatment can reduce gas emissions. The biogas that results from the anaerobic wastewater treatment may actually be harnessed and used as an alternative power source for cooking, lighting, heating and engine fuel.

II. Decide whether each of the following statements is true (T) or false (F) according to the text.
1～5: T F T F T

III. Translate the following words or phrases into English.

1. anaerobic wastewater treatment
2. biological agent
3. biodegradable material
4. anaerobic digestion
5. facultative and anaerobic bacteria
6. fatty acid
7. methanogenic bacteria
8. organic acid
9. carbon dioxide
10. greenhouse gases

IV. Translate the following sentences into Chinese.

1. 因为微生物在处理过程中所起的作用，厌氧废水处理又称为厌氧消化，从根本上来说，处理方式实际上是"消化"水中的污染部分。
2. 首先，产生水解脂肪、纤维素和蛋白质。由栖息细菌产生的细胞外酶将大分子水解产物分解成更小和更易消化的分子。
3. 接下来，脂肪酸被进一步分解，一些兼性厌氧菌进行这一分解过程。
4. 最后，产甲烷菌消化吸收这些脂肪酸，从而形成沼气。
5. 换句话说，收集和利用厌氧消化过程中产生的沼气和二氧化碳，这些生物气体就不会被排放至大气中。

V. Match the words and expressions with their meanings.
1～4: b a f d   5～8: h c g e

VI. Best choices.
1. A   2. B   3. D   4. D   5. D

VII. Fill in the following blanks.
图略。
Gas removal（气体导排）
Raw wastewater（原水）
Digesting tank (contents mixed) 消化池（物料混合）
Settler（沉淀池）
Solids return（污泥回流）

Part I 水处理

Unit 3 二级水处理

Text C 絮凝作用

悬浮在水中的天然砂颗粒很难清除，因为它们非常小，经常以胶质状态出现，自身带有的负电荷使得它们不可能结合在一起，形成容易去除的大颗粒。而沉淀可以清除这些颗粒，首先需要中和它们的电荷，其次使得这些颗粒彼此相互碰撞。这种电中和被称为凝聚，由小颗粒形成大絮体的过程称为絮凝。

絮凝是污水处理过程中的一个步骤。在絮凝池内，通过搅动或其他方式让水扰动起来，以便水中的颗粒能够相互碰撞，彼此吸附。最后，这些难以清除的水中小颗粒形成容易去除的絮体。在进入絮凝池的水里经常会加入一些化学品（最常见的是明矾），用以帮助大颗粒成型。

明矾（硫酸铝）是水处理中常见的三价阳离子来源。除其自身具有的高能正电荷之外，明矾还具备一个优势：一部分铝离子可通过化学反应变成氧化铝和氢氧化铝。

$$Al^{3+} + 3OH^- \rightarrow Al(OH)_3 \downarrow$$

这些复合物有黏性，质量大，不稳定的胶体粒子如果能够和絮凝物相接触，则可以极大地帮助沉淀池水的澄清。絮凝可以增强澄清效果。

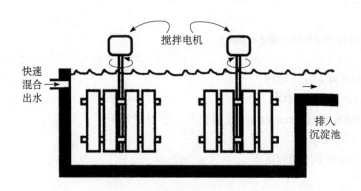

图 1 水处理过程中的絮凝器

絮状物一旦形成就必须从水中分离出来。这一过程总是在重力沉淀池中进行，重力沉降池能让比重大于水的颗粒沉淀到池底。沉淀池应设计为近似等速流及能够将涡流最小化，因此，设计沉淀池的两个关键因素为出入口结构。图2所示的为水处理沉淀池的一种出入口结构，此结构用于均匀分配进出水的流量。

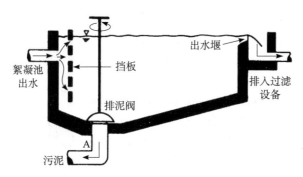

图2　水处理过程中的沉淀池

Key to the exercises

I. Answer the following questions after reading the text.
1. Because they are very small, often colloidal in size, and possess negative charges, and are thus prevented from coming together to form large particles that could more readily be settled out.
2. The removal of these particles by settling requires first that their charges be neutralized and second that the particles be encouraged to collide with each other.
3. In a flocculation tank, the water is stirred or otherwise moved around so that the particles move around, bump into other particles, and stick to one another. Eventually the small and difficult to remove particles in the water form large clumps which can then be easily removed.
4. Alum has an advantage in addition to its high positive charge: some fraction of the aluminum ions may form aluminum oxide and hydroxide by the reaction.
5. Settling tanks are designed to approximate uniform flow and to minimize turbulence.

II. Decide whether each of the following statements is true (T) or false (F) according to the text.
　　1～5: F F F T T

III. Translate the following words or phrases into English.
1. colloidal particle
2. negative charge
3. positive charge
4. charge neutralization
5. flocculation tank
6. aluminum sulfate
7. aluminum oxide
8. gravity settling
9. uniform flow
10. silt particle

IV. Translate the following sentences into Chinese.
1. 悬浮在水中的天然砂颗粒很难清除，因为它们非常小，经常以胶质状态出现，自身带有的负电荷使得它们不可能结合在一起，形成容易去除的大颗粒。
2. 而沉淀可以清除这些颗粒，首先需要中和它们的电荷，其次使得这些颗粒彼此相互碰撞。
3. 明矾（硫酸铝）是水处理中常见的三价阳离子来源。
4. 除其自身具有的高能正电荷之外，明矾还具备一个优势：通过化学反应一部分铝离子可能成为氧化铝和氢氧化铝。
5. 这一过程总是在重力沉淀池中进行，重力沉降池能让比重大于水的颗粒沉淀到池底。

V. Best choices.
1. B　2. A　3. C

VI. Fill in the following blanks.
图略。
Motors（搅拌电机）
To（settling tank）排入沉淀池
Influent from the（flocculator）絮凝池出水
挡板（Baffles）
出水堰（Weir）
Mud valve（排泥阀）
（Sludge）污泥

Part I 水处理

Unit 4 三级水处理

Text A 高级氧化技术处理废水的应用

概述

在废水处理过程中我们使用了很多物理、生物和化学的处理方法，但是普通的处理方法对于废水中的一些污染物在某种程度上也是无效的。化学氧化法是一种可以加强现有处理效果的一种处理方法。化学氧化法可以通过氧化和还原反应破坏某些复合物及其成分。

高级氧化法是产生羟基自由基的化学氧化，可以产生反应强烈的不稳定的氧化物，这种自由基需要现场产生，自由基在废水中接触到有机物时的反应器中产生。以下体系可产生羟基自由基：紫外线照射／过氧化氢，臭氧／过氧化氢，紫外线照射／臭氧，芬顿试剂（亚铁和过氧化氢），二氧化钛／紫外线照射等。

废水处理的应用

如图 1 所示，高级氧化处理可以在废水处理中用来：①减少有机物总量（化学需氧量）；②破坏特定污染物；③处理污泥；④不断增加不易分解有机物的生物利用度；⑤减少颜色和气味。

研究高级氧化处理是为了减少废水中总的有机物含量，这种有机物含量可通过测定生产清洁剂和地板护理产品的工业设备所产生的废水的化学需氧量 (COD) 来确定。这种废水包含了高达 5% 的表面活性剂、溶剂及螯合剂。生产设备产生的废水在排放到当地的公共处理设备前需要减少其化学需氧量浓度。

小型试验旨在评估用芬顿试剂减少废水中化学需氧量的可能性。实验的结果表明芬顿试剂很成功，正如图 2 所示，减少了 96% 以上的化学需氧量。在这些实验中我们还能发现该反应能释放出热量（减少每克的化学需氧量能释放 3.748 ± 0.332 焦的热量），这种热量可以供设备使用。

在其他处理完成后，高级氧化技术也可被用来去除经过处理还残留在废水中的特定的污染物，Bergendahl 等人使用芬顿氧化来减小水中有机污染物的浓度，在进行试验（见图 3）时，在芬顿氧化中出现的许多污染物被成功地降解。虽然没有三氯乙烷，在废水中的许多污染物浓度都明显地降低了（例如，间二甲苯和对二甲苯）。总的来说，经过芬顿氧化后的水中有机污染物减少了 81.8%，芬顿氧化对于水中甲基叔丁醚的矿化是非常有效的。

中川等人用试验规模反应器研究了臭氧对内分泌干扰物的破坏，也就是近期在废水中发现的污染物（见图 4），其中包括雌二醇、双酚 A、壬基酚浓度大幅降低，但是没有破坏雌激素酮。每升五毫克的臭氧量几乎可以完全破坏雌二醇、双酚 A、壬基酚，仅 20% 的雌激素酮得到减少。这些实验表明高级氧化的有效性取决于要降解的特定化合物。

高级氧化处理产生的羟基自由基对于处理和调节废水处理过程中产生的沉淀物也是非常有效的，因为它们能有效地破坏微生物的细胞壁。细胞物质在高级氧化时溶解，对于进一步的氧化和其他处理也有效，高级氧化技术能够进一步应用在废水技术中的两种工艺中（如图 5）。

图 5 中的处理方案（b）是由 Yasui 及其同事制定的，针对日处理量为 12 000 加仑的城市污水活性污泥系统，对部分回流污泥进行了臭氧化处理。该系统运行了 10 个月，没有剩余的污泥产生—— 这是无污泥系统。

我们把重点放在用其他技术整合高级氧化技术。分子筛沸石能大量吸收水中的有机污染物。然而，高级氧化能够破坏这些被吸附的污染物并且能再生吸附剂。图 6 表明，在水中硅质岩反复吸附氯仿后，达到硅质岩饱和并且不能再吸附任何污染物（循环 8 次后），但在高级氧化之后，硅质岩可重新获得原始吸收能力（循环 9 次后）。

结论

高级氧化处理可以被用在废水处理中，主要作用为：减少有机物总量、减少特定的污染物、处理污泥、增加不易分解有机物的生物利用度和减少颜色及气味。

Key to the exercises

I. Tell the differences between the two pictures.
此图为污泥处理中的实施高级氧化处理的两个方案：
(a) 厌氧消化前的污泥高级氧化；
(b) 再回收污泥高级氧化。

II. Choose the best choices to fill in the blanks:
1~5: C B D D A

III. Decide whether each of the following statements is true (T) or false (F) according to the text.
1~5: T F F T F

IV. Translate the following words or phrases into English.
1. hydroxyl radicals　　　　2. AOPs (advanced oxidation processes)

3. organic contaminants 4. wastewater treatment
5. Fenton's reagent 6. ozone 7. concentration
8. ultraviolet radiation 9. Adsorption cycles 10. tertiary treatment

V. Match the words and expressions with their meanings.
1~5: d b h f a   6~10: i g c j e

VI. Translate the following sentences into Chinese.
1. 这些实验表明高级氧化的有效性取决于要破坏的特定化合物。
2. 高级氧化法是产生羟基自由基的化学氧化，可以产生反应强烈的不稳定的氧化物。
3. 从生产设备中产生的废水中的化学需氧量在排放到当地的公共处理设备前需要降低它的浓度。
4. 高级氧化处理产生的羟基自由基对于处理和调节废水处理过程中产生的沉淀物也是非常有效的，因为它们能有效地破坏微生物的细胞壁。
5. 高级氧化处理在废水处理中的主要作用为：减少有机物总量，减少特定的污染物，处理污泥，增加不易分解有机物的生物利用度，减少颜色及气味。

Part I 水处理

Unit 4 三级水处理

Text B 活性炭法深度处理二级出水

简介

在过去的十年我们已经在关于废物处理和污染控制的观念上有一个根本的变化，最初，城市废物处理首要关注维护公共健康。然后我们更加关注城市的美观，例如消除可见的污染痕迹，维持水域海洋生物生存的氧气水平。在过去的十年中出现了两个新概念，第一是水已经变得稀缺所以我们要循环使用水，第二是要保持天然水体的纯净，这样水利用对于环境的影响可以减少到最小，这些新概念所要求的废水处理和我们曾经关注的公共健康、美观以及氧耗竭所引起的水处理有根本的不同。由此而论，活性炭吸附是废水处理的关键过程，它可以进行二级出水深度处理以满足当前的废水排放标准和接受水体质量的要求。

水处理中使用炭要追溯到 20 世纪。在 1883 年，据报道在美国有 22 家水处理厂使用了木炭过滤器。这种方法后来因为木炭吸附能力低而被放弃了。活性炭的使用源于 1913 年由维斯维克开始，然而，直到 1927 年，两家芝加哥肉类包装公司使用粉末活性炭来除去供水中的味道，出现了第一次有记录的城市水处理。到 20 世纪 30 年代，用粉末活性炭除去可溶性有机物味道和气味的方法得到迅速的推广。

1960 年，美国公共卫生署着手一项废水深度处理的研究项目，有两个既定目标：一是帮助减轻水污染问题，另一个是，使水能够直接使用并能循环使用。这个项目早期重点为吸附研究，这是达到既定目标的最有希望的方法；另一重点在研究最可行的吸附剂。公共卫生署开展了一系列的研究，1966 年，联邦水污染控制局评估了活性炭吸附对于废水处理的可行性。这些研究主要集中在两个方面：和最经济实用的炭吸附性能相关的炭的物质结构以及和炭的重新利用相关的炭的活化性能。

在这些研究的基础上，建立了几个示范工厂以获得商业设备的数据，其中一个工厂是洛杉矶县的县卫生设备区和联邦污水防治局共同合办的公司，位于加利福尼亚州的波莫纳市。这个工厂有 5 个碳处理器并且每分钟能生产 400 加仑水。第二家工厂位于加利福尼亚的太浩湖，是由南太浩湖公共设施区开办的，这个工厂的生产能力是每天 750 万加仑水。第三家工厂位于纽约长岛，本文将主要介绍。

背景

长岛的纳苏县占地 291 英里，与纽约市毗邻。在过去的二十年中，这个县经历了人口和水消耗的迅速增长。这个县的唯一水资源就是当地的地下水，其产量受限于其补给率,水资源耗费的连续增长性使水供应出现危机。过度抽取会导致地下水平面的降低以及含水层盐水的侵入。

区县的发展也减少了地下水的再次补充率,公共下水道系统的安装转移了以前通过化粪池和污水坑到海洋排水口再重新进入地下的废水。目前的推测表明,如果按照现在的趋势发展下去,从含水层取出的水的净含量到1977年将会超过地下水的再生量。

一个可以在许可范围内增加取水的计划是在含水层建造一个水力屏障。这个屏障会阻止含水层水的自然流出,据估计现在每天有3 000万加仑的水流失到海里去。它也会阻止盐水侵蚀到含水层,这早已是纳苏县一些地区的问题了。在纳苏县南部周边一系列的回灌井中注入三级处理废水就可以形成这道屏障。

水质量需求

为了提供一定量的水以供注入含水层的公共用水中,现在污水处理厂的污水必须接受额外的处理来满足下列需求:

①关于饮用水的美国公共卫生服务标准;

②注入系统的经济营运;

③和自然地下水的化学兼容性。

饮用水标准的采用最初是为了使公众在观念上接受把处理过的废水注入公共蓄水池中。现在的计划是在注入水和供水井之间保持至少间隔一英里,这个距离可以确保没有颗粒物或细菌会进入供水井。尽管如此,作为政策的关键就是注入水必须要符合饮用水的标准。

深度废水处理工艺

用来达到这些水质标准的深度废水处理流程包括明矾混凝、过滤活性炭吸附以及加氯消毒。

注入系统的经济运行方面的水质标准是示范项目的一部分,从开发至今的水注入系统来看,很明显颗粒物必须保持在最低水平。浊度应该要小于0.5JTU,在项目的早期阶段,溶解性气体应该是低水平的,但没有想象中的那么严重。超过1.0杰克逊单位的混浊度会导致压力的迅速累积,这样就导致注水需要在给定流速下进行。

兼容性最主要的问题是铁和磷酸盐的浓度。铁在蓄水层中沉淀,会形成不可消除的堵塞。磷酸盐的作用至今还未完全弄清,然而,观察到磷酸盐在注入水和注入水修复后浓度的改变,我们得到了磷酸盐和构成蓄水层的优质黏质沙土互相作用的结论。

海湾公园污水处理厂的终级沉淀池排出的污水被打入澄清池,在这里加入明矾和絮凝剂。污泥回流是为了促进絮凝和克服水质水量的突变。水流在重力的作用下进入两个平行运行的混合介质过滤器中,每个过滤器都包含了一个在12英寸沙层上的36英寸高的无烟煤床层。过滤回流是自动的,包括气体冲刷、表面冲刷和高低频率回流装置。

污水被泵入四个连续的颗粒活性炭吸附器。吸收管有序排列，可以通过旋转管道来控制流动方向并确保碳的有效利用。在碳接近耗竭时，可以通过液压将活性炭送至再生系统。失活的活性炭在多膛炉中通过可控制的燃烧，去除吸附在其中的有机物，恢复吸附活性。

用氯消毒再生水比把水抽到半英里外的测试注水现场要好。注水设施包括储水槽、消除余氯和溶解气体的脱气塔、注水和再生泵、注水井和12个观测井。这个注水井直径36英寸，深500英尺，里面装着一根18英寸的套管来支撑一个在高度420～480英尺的一个直径16英寸的网筛。网筛周围的空间填满了粒级砂和一个观测井，以及一个物探探头。其他的观测井位于注入井以上200英尺的地方。

Key to the exercises

I. Choose the best choices to fill in the blanks:

1～5: A C D A B

II. Decide whether each of the following statements is true (T) or false (F) according to the text.

1～5: T F T F T

III. Translate the following words or phrases into English.

1. aquifer   2. overpumping   3. drinking water   4. activated carbon
5. effluents   6. secondary effluent   7. marine life   8. adsorbant
9. public health   10. water plants

IV. Match the words and expressions with their meanings.

1～5: h a f c i   6～10: g j d b e

V. Translate the following sentences into Chinese.

1. 用来达到这些水质标准的深度废水处理流程包括明矾混凝、过滤、活性炭吸附以及加氯消毒。
2. 最初，城市废物的处理首要关注维护公共健康。
3. 水资源耗费的连续增长使水供应出现危机。
4. 从开发至今的水注入系统来看，较大颗粒物必须保持在最低水平。
5. 超过1.0杰克逊单位的混浊度会导致压力的迅速累积，这样就导致注水需要在给定流速下进行。

VI. Supplementary reading comprehension.

Decide whether each of the following statements is true (T) or false (F) according to the text.

1～5: F T F F T

活性炭吸附系统

废水处理的碳吸附系统设计要考虑以下几个参数：
- 炭的种类——颗粒状还是粉末状。
- 物理结构——向上流动还是向下流动，还是混合流，平行还是连续，填充层还是扩张层，外部再生流还是连续流。
- 炭的性能——停留时间，剂量率。
- 操作方法——纯吸附，过滤，生物化学处理方法。

就纳苏县的项目而言，因为目前炭再生的工艺水平，选择颗粒状的炭而不是粉末状的炭，其实粉末状炭在某些方面比颗粒状炭有优势。

粉末炭原始成本比较低，每磅只要花 7.5 美分，而颗粒炭每磅要花 30 美分。粉末炭反应得更快更完全，它的剂量也可以随着系统水流中成分的改变而变化。从另一方面而言，即使粉末炭比较便宜，也不能用了一次后就丢弃。粉末活性炭的再生试验工作现在正在进行，但是还没到可以在示范工厂大规模使用的阶段。废弃的粉末炭最可行的处置方法是脱水和焚化。

颗粒炭用于工业已经很多年，再生技术也十分完善。它在操作上还具有粉末炭不能相比的额外运行安全技术。废水处理时进水成分常常改变，粉末碳的剂量不能适应这些变化，出水水质会反映粉末炭剂量的不足。颗粒炭能够经受水质成分的大幅度改变以及稳定出水水质。这一点以及活性炭的再生有效性是海湾公园项目中选择颗粒炭的主要因素。

即使选择了颗粒炭或者粉末炭之后，我们还要进行更进一步的选择。活性炭是由多种原材料制成的，例如煤、木材、坚果壳以及纸浆废物。炭必须能在经受多次再生后保持最小程度的退化和磨损。因为从煤中提取的活性炭比其他的炭更坚硬，密度更大，这种炭被特定用在海湾公园项目中。

工艺试验的结果表明炭可以另外加入含量不高于 0.5% 的铁，铁的限量使炭在等级上发生了变化，所以炭的规格可以满足商业产品的需要，原始炭的大小是 8×30（通过 8 号网筛，但是被截留在 30 号网筛上），再生炭的大小是 14×40。

活性炭吸附系统有很多类型，其中包括上向流扩张床、上向流压缩床、下向流单极、下向流多级以及准逆流系统。在第一个装置中水流上升，在第二个装置水流下降，快要耗尽的炭渐渐从第一个装置中移除，并逐渐在第二个装置中再生或者补充。

上向流系统的优点是不太容易堵塞更易于持续逆向运行，这在理论上能使炭得到最有效利用。下向流系统定期逆流以免形成水头损失和因多级操作而导致倒流运行。与再生和补充的花费相比，设备的费用是次要的。下向流系统就流量而言更简单更灵活。纳苏县的项目选择了一个四级下向流系统，四个容器并联放好，每一个单元都能作为第一个单元。在常规操作中，这个流程适用于含有的活性炭接近耗竭的容器。当水流从一个单元到另一个单元时，它会连续碰到更多的活性炭，直到在最后一个单元中，它通过了刚再生过的活性炭。

当出厂水中有机物的含量开始超出预期标准时，第一个单元就会离线，单元中的炭经液压传到炭再生系统中的脱水槽中去。转换一完成，储蓄槽中的再生炭就被送到单元中。这时单元就会重新回到线路的末端。通过这种方式来保持运行的逆流。

实验室和中试工厂的研究可提供其他的设计数据，实验室的研究过去常用于生成吸附等温线，用来表征炭的吸附容量。柱层测试常用来确定所需要的接触时间。水力负荷变得方便设备的设计。就纳苏县项目而言，以下采取的设计参数都是基于一年的中试工厂操作。

- 总接触时间（空床体积）24 min。
- 水力负荷（行进流速）7.5 gpm/sqft。

综合这些因素，每一个容器的直径是 8 英尺，深 6 英尺。每一个容器能容纳 300 立方英尺或者大概 9000 磅炭。耗气率大概是每磅 800 加仑或者每需炭 1.25 磅 1000 加仑。

经济

深度废水处理过程的单元费用如下表所示。这个表是基于二级污水中化学需氧量去除率 90%，从 50 mg/L 到出厂水的 5 mg/L，磷酸盐去除率 90%，从 30 mg/L（例如磷酸根）减少到 3 mg/L。

每年的费用估计是全部资本的 8.5%，其中包括债务还本付息和维护、维修、更换费用。单位成本也包括设计能力的连续操作（100% 的负荷系数）。这个费用只包括治理费用，不包括传输和注入设备。

结论

纳苏县项目说明了使用物理化学方法处理二级出水的可行性，其中生化处理工艺包括了活性炭、除去持久性有机物以及生物处理。出水满足美国公共卫生署标准对饮用水的要求，满足民用给水对浊度、颜色、气味等水质标准要求。出水可以直接注入地下而不引起任何蓄水层的退化。在目前进行的实验的基础上，可以确定使用处理过的废水作为水利屏障来防止海水的入侵在技术上是可行的。

这个项目展示了淡水供应紧缺区域的废水利用可能性，在有公用水和排污系统的地方废水总是可用的。新采用的水质标准要求更多的社区能提供高于常规二级处理方法的废水处理工艺。更多形式的高质量再生水可以使用活性炭处理后的出水。

Part I 水处理

Unit 5 废水处理平台

Text A 污水处理平台的样板工程（1）

污水处理是指去除如地表径流、家庭和商业废水这样的废水和生活污水中污染物的总流程。它包含去除物理、化学和生物污染物的与之相对应的各个物理、化学和生物流程。其目的在于通过处理后产生对环境安全的适用于进一步处置和再利用的废水（处理过的出水）和固体废弃物（处理过的淤泥）。

人们可以在产生污水的地方就近进行污水处理，或者把污水通过管道和泵站网络来收集并运送到城市污水净化厂进行处理。

污水处理通常包括三个阶段，即初级、二级和三级处理。

初级处理包括一个暂时存储污水的沉淀池，可使较沉的固体物质沉到池底，而油、油脂和更轻的固体物质则浮于污水表面。去除沉淀物和漂浮物后，就可将剩余的液体排放至二级处理设施。

二级处理是要去除溶解的污染物质和悬浮生物物质。它通常通过在一定可控区域内的以水为载体的微生物来进行。二级处理也许需要一个分离过程来去除被处理过的将要排放出去或要进入三级处理的废水中的微生物。

三级处理有时候会被我们定义为在初级处理和二级处理后继续截留废水进行深度处理后才使之进入极度敏感脆弱的生态系统的所有的废水处理过程。有时候处理过的水会在排放到溪流、河流、海湾、潟湖或湿地之前进行化学或物理消毒。如果处理过的水足够干净，则可以用于补充地下水或农业用途。

长沙环境保护职业技术学院的污水处理平台是一个水处理的样板工程。2002年由该校的环境工程系设计和建立并用于培训学生技能。在此三级递进的废水处理系统中采用了生物接触氧化、溶气浮选和絮凝这三种广泛应用的废水处理技术。

从学校的第三教学楼和办公楼流出来的废水可以作为原水收集起来，然后运输到这个污水处理平台。

原水水质数据如下：

表1 原水水质（mg/L）

化学需氧量	5天生化需氧量	悬浮物	总氮	总磷
400	200	220	40	8

设计设备容量：废水流量 8 m^3/h。

设计出水水质数据：

表2 出水水质（mg/L）

项目	化学需氧量	5天生化需氧量	悬浮物	原油	细菌总数（个/L）
最大值	16.8	2.8	17.2	0.3	<10
最小值	8.4	1.4	7.6	0.2	<10
平均值	12.5	2.2	10.3	0.25	<10
国家一级排放标准	100	30	70	10	—
其他城市水质标准 GB/T 18920—2002	50	10	10	—	100
处理效率/%	96.4	98.5	98.4	96.6	99.9

工艺流程：这个废水处理系统采用多级（分级）和平行（并联）连接。转换阀和出水管是被留作随机掌控操作流程转换的。

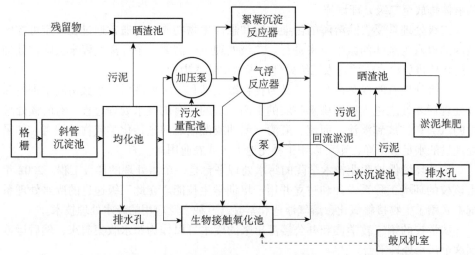

图1 长沙环境保护职业技术学院污水处理厂的工艺流程

格栅（格栅槽）：

当污水流经条形格栅时，格栅会滤除污水流中携带的所有粗大的物体，如易拉罐、破布、树枝和塑料袋等。在为大量人口服务的现代化的污水处理厂，最常使用的是自动化的斜耙格栅，而像我校的这种规模较小的不那么现代化的污水处理系统使用手动净化格栅即可。机械式条形格栅的除污动作频率通常是根据其上的积累物和污水流速确定的。把滤除的固体物质收集起来后，可对固体进行填埋处理或焚烧。人们用不同尺寸的格栅或筛网来优化固体物质的去除。如果不移除那些显而易见的固体，它们将会被夹带在管道中，成为污水处理厂中运转流程中的一部分，且可能导致对设备的重大损害，或使整个处理过程低效率运作。

沙砾与油脂去除（斜管沉淀池）：

在校园污水处理系统的斜管沉淀池中，污水流速的调整是根据沉淀的砂、沙砾、石头和碎玻璃的量而进行的。因可能会损坏泵或其他设备，这些颗粒物将会被移除。对于小的污水处理系统，并没必要采用沉砂池，不过大一些的污水处理厂就需采用沉砂池了。

在我们校园的污水处理系统中，当污水通过斜管沉淀池时，撇油器将会收集浮于水面的脂肪，从而将脂肪和油脂从水中去除。在现代污水处理厂中，位于池底部的鼓风机可用来促使脂肪变成泡沫状。然而，许多处理厂是使用配有机械表面撇油器的初级澄清池来去除水中的沙砾与油脂的。

沉砂池有3种类型：平流沉砂池、竖流沉砂池和径向流沉砂池。

均流：

均流池是白天或水流量达到峰值时一个临时存储水的地方。它既可在水处理平台维修时临时保存注入的污水，也是一种过滤和分配有毒或含有高浓度废弃物废水的排放批次的工具，否则这些过量的有毒物质或高强度污染物会抑制二级生物处理。

在其他的污水处理厂中，均流池还需要配备能进行可变流量控制的装置，通常包括分流装置、清洗装置，也许还包括曝气机。如果将均流池置于格栅和去除沙砾的流程之后，则污水的净化可能更简单一些。

Key to the exercises

I. Best choices.

1. D 2. B 3. C 4. A 5. C

II. Translate the following words or phrases into English.

1. sewage treatment plant   2. drain-outlet   3. skimmer
4. discharging standard   5. return sludge 6. aerator   7. bar screen
8. grill trough   9. disinfect   10. grease removal

III. Match the words and expressions with their meanings.

1~5: j i k a c   6~10: d e f g b

IV. Match the words and expressions with their meanings.
1~5: b d a e c

V. Translate the following sentences into Chinese.
1. 沉淀物和漂浮物被去除后，废水可以排出以进行二级处理。
2. 在最后排放或者三级处理之前，二级处理工程中也许需要分离过程来去除一级处理后的废水中的微生物。
3. 流入的污水流经一个格栅，一边去除这些污水所携带的易拉罐、破布、树枝和塑料袋等所有大的物体。
4. 如果没有去除粗大的固体物质，它们就会被水流挟带进入水处理平台的管道和运转部件，有可能损坏水处理设备，使水处理过程失效。
5. 对于小型的污水处理系统，也许用不着配备沉砂池，但在大型的污水处理厂，粗砂去除是很有必要的。

VI. Make oral presentation on the process flows of the sewage treatment plant in Changsha Environmental Protection Vocational College.
Open.

Part I 水处理

Unit 5 废水处理平台

Text B 水处理平台的样板工程（2）

二级处理（生物接触氧化池、絮凝沉淀反应器、气浮反应器）：

二级处理旨在大幅度降低来源于人类垃圾、食物垃圾、肥皂水和洗涤剂废水中的微生物。城市中大多数污水厂是利用好氧生物处理法来处理静止的污水。

（1）生物接触氧化是一种常用的二级水处理工艺，它通过为废水中的细菌和微生物提供生长所需的养分来分解和抑制废水中的有机污染物的生长。为了使水处理有效进行，需提供生物生存所需的氧气和食物。细菌和原生动物消耗可生物降解的可溶有机污染物，并把相当多的不那么易于溶解的部分物质凝聚成絮状物。在生物接触氧化池底部的鼓风机可用来帮助好氧生物提供其所需的溶解氧。

（2）絮凝沉淀法则广泛应用于饮用水的净化以及污水、雨水和其他工业废水流的处理。

在絮凝沉淀反应器中，由于澄清剂的添加，在絮状和小薄片状的悬浮物中会形成胶粒物质，并会下沉到反应器的底部。

这个操作与沉淀的不同点在于：在絮凝之前，胶粒物质仅仅只是悬浮在液体中，而不是真正地溶解于溶液里。絮凝系统没有形成块状物体是因为所有絮凝物都处于悬浮状态。

（3）溶气浮选法也是一种水处理方法，即通过移除如油或固体等悬浮物来澄清废水（或其他水）。为达到移除悬浮物的目的，我们可将水中的空气溶解或使絮凝箱或絮凝池的废水处于压力下，然后再将其置于大气压下释放空气。释放出来的空气可形成细小的泡沫黏附在悬浮物质上，使得这些悬浮物质漂浮到水面，然后就可以被撇沫装置去除。溶气浮选法广泛应用于炼油厂、石油化学产品和化工厂、天然气加工厂和造纸厂的工业污水废水处理。

二次沉降：

二级处理阶段的最后一步是沉淀生物絮状物或使过滤物体通过二次澄清器，进而让处理后的污水中的有机物和悬浮物处于低含量水平。

污泥的处理和清除：

在水处理过程中积累的污泥必须要用一个安全有效的方式来处置。分解的目的是减少固体颗粒物中有机物和致病微生物的数量。最常见的处理方法包括厌氧分解、好氧分解和堆肥。

污泥处理取决于固体的生成数量和其他具体的场地条件。堆肥法最常用于小规模的污水处理厂，小型的操作采用好氧消化，大规模的操作就采用厌氧消化。

主要设备和结构：

表 1　主要设备和结构

	设备名称	数量	规格
1	格栅槽	1	2170mm×420mm×600mm
2	斜管沉淀池	1	5000mm×1100mm×3200mm
3	废水流均化池	1	6400mm×3200mm×4800mm
4	絮凝沉淀反应器	1	
5	气浮反应器	1	气浮溶气罐：φ600×800 反应器：φ900×800 (Separate part); φ300×3300 (Reaction part)
6	生物接触氧化池	2	2600mm×2200mm×5000mm
7	二次沉淀池	1	3300mm×1800mm×3600mm
8	鼓风机室	1	6000mm×6000mm JTS-50 blower×2
9	水泵	8	IHG20-110、IHG25-110、IHG40-125 & 32WQ8-12-0.75
10	污泥干化池	1	7960mm×3000mm×1300mm

Key to the exercises

I. Decide whether each of the following statements is true (T) or false (F) according to the text.

1～5: T T F F F

II. Translate the following words or phrases into Chinese.

1. 生物接触氧化池　2. 澄清剂　　　3. 絮凝沉淀反应器　4. 格栅槽
5. 二次沉淀池　　　6. 鼓风机室　　7. 均化池　　　　　8. 沉淀池
9. 致病微生物　　　10. 过滤材料

III. Match the words and expressions with their meanings.

1～5: d j i h f;　6～10: g e a b c

IV. Match the words and expressions with their meanings.
1～5: c e a b d

V. Translate the following sentences into Chinese.
1. 大部分城市污水处理厂采用好氧生物处理工艺来处理沉淀后的污水。
2. 生物接触氧化池底部的鼓风机可用来给好氧微生物提供溶解氧。
3. 溶气浮选法是一种通过去除油或固体物质来澄清废水（或其他水）的水处理工艺。
4. 最常用的水处理方法包括厌氧消化、好氧消化和堆肥。
5. 污泥处理方法的选用要根据固体物质的产生量和其他具体的场地条件。

VI. Tell the names and amount of the devices in the table in the text.
Ommited.

Part II 空气污染控制技术

Unit 1 大气污染控制（气体）

Text 气体的控制

空气是一种重要的自然资源，是地球生物生存的必需品。大气中的空气为一切动植物提供其赖以生存的氧气，因此，给一切的生物活动提供优质的空气是极其重要的。然而，社会的工业化、农业的集约化、机动车的引进以及人口剧增带来了大规模污染，保证良好的空气质量变得日益困难。上述的人类活动会产生初次污染和二次污染，极大地改变了空气的组成成分，因此，Kaifu 将空气污染定义为由化学物、颗粒污染物（PM）或者生物材料的引入而给人类或者其他生物有机体造成伤害或不适，或者给自然环境或建筑环境造成损害。

为了控制气态标准污染物、挥发性有机化合物（VOC）及其他气态有毒物质，ICMA 已确定通过利用吸收、吸附、焚烧三种基本技术来控制气态污染物。这些技术将根据污染物的类型来确定是单独使用还是组合使用，下面是这三种技术的详细说明。

吸收

在空气污染控制方面，吸收涉及气态污染物由气态转变成接触液体（比如水）。液体必须是能够溶解污染物或与之发生化学反应。与控制悬浮颗粒装置功能类似的湿式的除尘器可用于气体吸收，这种气体吸收也可以在填料洗涤器或塔内进行，其中的液体存于一个潮湿的表面上，而不是像水滴一样悬浮在空中。

吸附

与吸收相比，气体吸附是一种表面现象。气体分子被吸附并固定在固体表面上。气体吸附方法用于在一些挥发性溶剂（如苯）的恢复和工业设施中的挥发性有机化合物的控制过程中对各种类型和阶段的化工生产和食品加工进行气味控制。

活性炭（热成炭）是最常用的吸附材料之一，这种材料多孔，具有极高的表面吸附率。活性炭作为一种吸附剂在净化包含挥发性有机化合物空气流、溶剂的回收以及气味控制方面尤其有用。使用设计合理的炭吸附装置，清洁气体的效率可超过95%。

焚烧

焚烧或燃烧可以将挥发性有机化合物和其他气态烃污染物转化为二氧化碳和水。挥发性有机化合物和烃废气通常在特殊的焚化炉即后燃室进行。要实现完全燃烧，后燃室必须提供适量的搅动和燃烧时间，并且必须保持足够高的温度。充分搅动或混合是燃烧的关键因素,因为它减少了所需的燃烧时间,降低了燃烧温度。当废气本身就是一种可燃混合物且不需要添加空气或燃料时,便可使用所谓的"火焰直接焚烧法"。

Key to the exercises

I. Answer the following questions after reading the text.
1. It is because of the industrialization of society, intensification of agriculture, introduction of motorized vehicles and explosion of the population.
2. People's activities generate two types of air pollutant: primary and secondary air pollutants.
3. Kaifu therefore defined air pollution as the introduction of chemicals, particulate matter (PM) or biological materials that cause harm or discomfort to humans or other living organisms, or cause damage to the natural environment or built environment into the atmosphere.
4. The three basic techniques employed to control gaseous pollutants are absorption, adsorption, and incineration.
5. Wet scrubbers, and packed scrubbers or towers can be used for gas absorption.

II. Decide whether each of the following statements is true (T) or false (F) according to the text.
1～5: T F F F T

III. Translate the following words or phrases into English.

1. primary air pollutants   2. secondary air pollutants
3. particulate matter   4. gaseous criteria pollutants
5. packed scrubber   6. activated carbon
7. afterburner   8. volatile organic compounds (VOCs)
9. odour control   10. spray head

IV. Translate the following sentences into Chinese.
1. 然而,社会的工业化、农业的集约化、机动车的引进以及人口剧增带来了大规模污染,保证良好的空气质量变得日益困难。
2. 在空气污染控制方面,吸收涉及污染物由气态变成接触液体(比如水)的转化。液体必须是能够充当污染物的一种溶剂或通过化学反应捕获它。

3. 活性炭作为一种吸附剂在净化包含挥发性有机化合物的空气流、溶剂的回收以及气味控制方面尤其有用。
4. 焚烧或燃烧可以将挥发性有机化合物和其他气态烃污染物转化为二氧化碳和水。
5. 要实现完全燃烧,后燃室必须提供适量的搅动和燃烧时间,并且必须保持足够高的温度。

V. Match the words and expressions with their meanings.
1～5: g d a j h   6～10: i c f b e

VI. Best choice.
1～5: C C B A D

VII. Fill in the following blanks.
图略。
清洁空气出口(Clean gas)out   喷头(Spray heads)
水进口(Water)in   污染空气进口(Contaminated gas)in
污染物出口(Contaminants)out

Part II 大气污染控制技术

Unit 2 大气污染控制（颗粒物）

Text 微粒的控制

尘粒可以通过各种物理过程将其从污染气流中去除。细粒子收集设备的常见类型包括旋风分离器、洗涤器、静电除尘器、袋式除尘器等。一旦收集到颗粒物，它们彼此黏附在一起成团，以便随时可以从设备中清除或者将其在垃圾填埋场处理掉。静电除尘器和袋式除尘器常常在发电厂使用，旋风分离器、洗涤器、静电除尘器、袋式除尘器作为空气清洁设备的工作原理由 ICMA 描述展示如下。

旋风分离器

旋风（见图1）通过促使脏气流在圆柱腔内进行螺旋式流动来清除微粒。脏空气在旋风分离器的外壁沿切线方向进入圆柱腔，在圆柱腔内打旋的脏气流形成旋涡，较大的颗粒物由于惯性较大而向外移动，并被迫旋转至圆柱腔壁。大颗粒由于跟腔壁摩擦而速度减慢，并滑落下来在旋风分离器的底部形成一个锥形灰斗。清洁的空气在内筒里呈较细的螺旋形向上旋转，并从顶部的出口涌出，聚集的颗粒粉尘可从漏斗中定期移除并进行处置。旋风主要用于去除相对较粗的颗粒物。对于直径大于 $20\mu m$（0.0008英寸）的颗粒，它们可以达到 90% 的清除效率。然而，旋风并不足以满足严格的空气质量标准。它们通常用作空气预滤器，然后再跟效率更高的空气清洁设备如静电除尘器、袋式除尘器等一起使用。

湿式除尘器

湿式除尘器通过直接接触喷雾剂如水或其他液体来捕获悬浮颗粒。实际上，脏气流中的颗粒碰到喷雾中的无数小液滴并黏附其上，洗涤器即在这一过程中将脏气流中的颗粒清除掉。

静电除尘器

静电除尘是一种去除脏气流中细粒子的常用方法。在一个静电除尘器中（见图2），脏气流中的悬浮颗粒在进入该装置时便携带电荷并在电场的作用下被清除。除尘装置包括气流分布挡板、放电和收集电极、灰尘清理系统以及收集漏斗。高直流电压（高达 100 000V）作用于放电电极以给颗粒充电，带电颗粒即被吸引至相反电荷收集电极而被捕获。

在一个典型的静电除尘装置里，收集电极由一组大型的矩形金属板组成，这些金属板垂直地分布在一个结构箱里并相互平行。这种金属板往往有数百块，其结合表面面积达数万平方米。无数条放电极导线悬挂在集合板之间而带负电，而板材因接地而带正电。

袋式除尘器

效率最高的去除悬浮颗粒的一个装置类似于一个织物过滤袋，俗称袋式除尘。一个典型的袋式除尘器（见图3）由一列列长而窄的袋子组成——每个袋子直径约25 cm（10英寸）——所有袋子被倒立地悬挂在一个大机箱里。含尘气体在风扇的作用下通过机箱底部被向上吹起。粒子被过滤袋吸附住，而清洁的空气穿过织物从顶部出口出来。

在此讨论的微粒控制技术都有利有弊。静电除尘器可以在低压条件下处理大体积流量的空气并且可达到非常高的效率（99.9%）。它们的成本大致等同于织物过滤器，但对工艺操作条件的变化灵活性相对较差。湿式除尘器也可以实现高效率，其最主要的优势在于在去除微粒的同时可去除部分气态污染物。然而，它们只能处理较小的气流量的气流（最大 3000 m³/min），由于高压因素，其运作成本可能会很高，并产生湿泥，处理困难。对于较高的废气流速且要达到对颗粒超过99%的清除率，设备可以选择在成本费用方面相差无几的静电除尘器和织物过滤器。

建议

为有效控制工业应用中的 PM_{10}，建议使用静电除尘器和袋式除尘器。这些除尘器应该要按照它们的设计效率来操作。如果没有具体的排放要求，应要达到小于50 mg/m³ 的排放量，最大不能超过 50 mg/m³。

对于含有可溶性有毒物质以及流量小于 3000 m³/min 的气体，可使用湿式除尘器。旋风和机械分离器应只能用作在袋式除尘器或静电除尘器之前的预洗设备。

Key to the exercises

I. Answer the following questions after reading the text.

1. The common types of equipment for collecting fine particulates are cyclones, scrubbers, electrostatic precipitators and bag house filters.
2. Electrostatic precipitators and fabric-filter bag houses are often used at power plants.
3. The cleaned air emerges from an outlet at the top and accumulated particulate dust comes out from the hopper for disposal.
4. Because the particles and the collection electrodes are oppositely charged.
5. The precipitation unit comprises baffles for distributing airflow, discharge and collection electrodes, a dust clean-out system and collection hoppers.

附　录　191

II. Decide whether each of the following statements is true (T) or false (F) according to the text.
1～5：F F T F F

III. Translate the following words or phrases into English.
1. suspended particles/particulates　　2. cyclone
3. electrostatic precipitator　　4. wet scrubber
5. bag house filter　　6. electric charge
7. electric field　　8. DC voltage
9. fabric-filter bag　　10. conical dust hopper

IV. Translate the following sentences into Chinese.
1. 旋风分离器通过促使脏气流在圆柱腔内进行螺旋式流动来清除微粒。
2. 实际上，脏气流中的颗粒碰到喷雾中的无数小液滴并黏附其上，洗涤器即在这一过程中将脏气流中的颗粒清除掉。
3. 效率最高的去除悬浮颗粒的一个装置类似于一个织物过滤袋，俗称袋式除尘器。
4. 一个典型的袋式除尘器由一列列长而窄的袋子组成——每个袋子直径约 25 cm（10 英寸）——所有袋子被倒立地悬挂在一个大机箱里。

V. Match the words and expressions with their meanings.
1～5: e f a h j　　6～10: b c d g i

VI. Best choice.
1～5: C A C A D

VII. Fill in the following blanks.
图略。
清洁空气出口（Clean gas）out
含尘空气进口（Particle-laden gas）in
空气流旋转使颗粒向外靠至内壁并最终落向旋风基底　Spinning gas stream forces particles to (outside walls) and then to the (base) of the cyclone
收集到的颗粒物（Particles）collected

正电集流板（Positively）charged collector plates
清洁空气流 Clean（gas stream）
负电金属网格（Negatively）charged metal grid
清除的颗粒（Particles removed）
含颗粒的空气流带上正电 Gas stream containing（particles）picks up（negative charge）

清洁空气出口（Clean gas）out
过滤袋（Bag filters）
含尘空气进口（Particle-laden gas）in

VIII. Work in pairs or groups and list the principal advantages and disadvantages of the particulate control technologies discussed in this unit.

Advantages	Disadvantages
Inertial or impingement (cyclone) separators • Low capital cost (approximately US$1/cu ft/min flow rate) • Relative simplicity and few maintenance problems • Relatively low operating pressure drop (for the degree of particulate removal obtained) in the range of approximately 5-15cm (2-6 inches) water column • Temperature and pressure limitations imposed only by the materials of construction used • Dry collection and disposal • Relatively small space requirements	• Relatively low overall particulate collection efficiencies, especially for particulate sizes below 10 mm • Inability to handle sticky materials
Wet scrubbers • No secondary dust sources • Relatively small space requirement • Ability to collect gases, as well as particulates(especially "sticky" ones) • Ability to handle high-temperature, high-humidity gas streams	• Potential water disposal/effluent treatment problem • Corrosion problems (more severe than with dry systems) • Potentially objectionable steam plume opacity or droplet entrainment

续表

Advantages	Disadvantages
• Low capital cost (if wastewater treatment system is not required) • Insignificant pressure-drop concerns for processes where the gas stream is already at high pressure • High collection efficiency of fine particulates (albeit at the expense of pressure drop)	• Potentially high pressure drop-approximately 25 centimeters (10 inches) water column and horsepower requirements • Potential problem of solid buildup at the wet-dry interface • Relatively high maintenance costs
Electrostatic precipitators • Collection efficiencies of 99.9% or greater of coarse and fine particulates at relatively low energy consumption • Dry collection and disposal of dust • Low pressure drop-typically less than 1-2 cm (0.5 inch) water column • Continuous operation with minimum maintenance Relatively low operation costs • Operation capability at high temperatures (up to 700℃, or (1,300°F) and high pressure (up to 10 atmospheres, or 150 pounds per square inch, psi) or under vacuum • Capability to handle relatively large gas-flow rates(on the order of 50,000m^3/min)	• High capital cost-approximately US\$160/square meter (\$15/square foot) of plate area • High sensitivity to fluctuations in gas stream conditions (flow rates, temperature, particulate and gas composition, and particulate loadings) • Difficulties with the collection of particles with extremely high or low resistivity • Relatively large space requirement for installation • Explosion hazard when dealing with combustible gases or particulates • Special precautionary requirements of safeguarding personnel form high voltage during ESP maintenance by deenergizing equipment before work commencement • Production of ozone by the negatively charged electrodes during gas ionization • Highly trained maintenance personnel required

续表

Advantages	Disadvantages
Fabric filter systems (baghouses) • Very high collection efficiency (99.9%) for both coarse and fine particulates • Relative insensitivity to gas stream fluctuations and large changes in inlet dust loadings (for continuously cleaned filters) • Recirculation of filter outlet air • Dry recovery of collected material for subsequent processing and disposal • No corrosion problems • Simple maintenance, flammable dust collection in the absence of high voltage • High collection efficiency of submicron smoke and gaseous contaminates through the use of selected fibrous or granular filter aids • Various configurations and dimensions of filter collectors • Relatively simple operation	• Requirement of costly refractory mineral or metallic fabric at temperatures in excess of 290℃ (550 °F) • Need for fabric treatment to remove collected dust and reduce seepage of certain dusts • Relatively high maintenance requirements • Explosion and fire hazard of certain dusts at concentration (~50 g/m^3) in the presence of accidental spark or flame, and fabric fire hazard in case of readily oxidizable dust collection • Shortened fabric life at elevated temperatures and in the presence of acid or alkaline particulate or gas constituents Potential crusty caking or plugging of the fabric, or need for special additives due to hygroscopic materials, moisture condensation, or tarry adhesive components • Respiratory protection requirement for fabric replacement • Medium pressure-drop requirements-typically in the range of 10-25centimeters (4-10 inches) in water column

IX. Reading Material.

Decide whether each of the following statements is true (T) or false (F) according to the given materials.

1~5: F T F F T   6~10: T T F T T

大气颗粒物：污染预防和控制

大气颗粒物（PM）的排放可通过实施污染预防和排放控制措施实现减量。应该强调的是预防，因为这往往比控制更具成本效益。应特别重视在与颗粒物排放有关的有毒物区域内的污染治理措施，因为这种有毒物质可能会给环境造成重大的风险。

污染预防办法

（1）管理

改进的工序设计、操作、维修、内务管理和其他管理措施可以减少排放量。通过提高燃烧效率，可以极大地减少不完全燃烧的产物（颗粒物的成分之一）。适当的燃料焚烧法与燃烧区配置，以及足够过量的空气，可产生更少的不完全燃烧产物。

（2）燃料的选择

选择较清洁的燃料可减少大气颗粒物的排放。天然气燃烧排放出的颗粒物几乎可以忽略不计。石油的燃烧所排放出的颗粒物比燃煤少得多。低灰分化石燃料包含较少的不燃、成灰矿物，因而生成颗粒物的水平也较低。以油为基础的轻馏分燃烧比重渣油燃烧排放的颗粒物水平低。然而，燃料的选择通常既要考虑经济也要考虑环保。

（3）清洁燃料

清洁燃料使灰烬减少，这就降低了颗粒物的产生和排放量。通过洗涤和选矿让煤得到物理清洗，这可以减少其灰烬和硫的含量，但必须要小心地处理这一清洗过程中所产生的大量固体废弃物和废水。另一种可替代煤炭清洁的方法便是灰分较高和较低的煤掺烧。除了减少微粒的排放量，低灰煤也有助于维持更好的锅炉性能并减少锅炉的维护成本和停机时间，从而节省部分煤炭清洁费用。例如，对于东亚地区的一个项目来说，对煤清洁的投资可有26%的内部收益率。

（4）技术和工序的选择

使用更有效的技术或工序可以减少PIC的排放。先进的煤炭燃烧技术，如煤炭气化和流化床燃烧就可将PICs降低约10%。封闭式煤炭破碎机和磨机排放较少的颗粒物。

排放控制方法

可使用各种具有不同物理和经济特点的颗粒清除技术。

撞击式分离器依靠粒子的惯性特点将它们从载气流中分开。撞击式分离器主要用于收集中等大小和较粗的颗粒。它们包括沉降室和离心旋风（直流或更常用的反流旋风）。

静电除尘器（ESPs）通过使用静电场将粒子吸引至电极上来去除粒子。

过滤器和集尘器（袋）使烟气通过一个充当过滤器的织物来收集灰尘。最常用的是袋式过滤器或袋式除尘器。各种类型的过滤介质包括织物、针刺毛毡、塑料、陶瓷以及金属。

湿式除尘器依靠液体喷雾从气体流中去除尘粒。它们主要是以微粒控制作为辅助功能来去除气态污染物。

设备的选择

颗粒排放控制设备的选择受到环境、经济和工程因素的影响：

环境因素包括：(a) 控制技术对周围空气质量的影响；(b) 污染控制系统对废水和固体废物的产出量和性能的作用；(c) 可允许的最大排放要求。

经济因素包括：(a) 控制技术的资金成本；(b) 技术的营运和维修成本；(c) 设备的预期使用寿命和挽救价值。

工程因素包括：(a) 污染物特征如物理和化学属性——浓度、颗粒形状、大小分布、化学活性、腐蚀性、耐磨性和毒性；(b) 气体流的特性，如体积流量、含尘量、温度、压力、湿度、组成、黏度、密度、反应性、可燃性、腐蚀性和毒性；(c) 控制系统的设计和性能特征，如对降压、可信度、可靠性、实用性和维护要求、温度限值，以及颗粒物的大小、重量、分级效率和气体或蒸汽的传质或污染物的破坏能力。

Part III 固体废物管理

Unit 1 固体废物管理概述

Text 固体废物管理

人类总是产生垃圾并总是以某种方式去处理这些东西，因此固体废物管理并不是个新问题。改变了的只是产生的废物的类型和数量，处理废物的方法和人们对该怎样处理废物的价值观念。

过去，垃圾总是以最便利的方式被丢弃，很少考虑其对人类健康和环境的影响。现代卫生学观念还未形成之前，城市街道就像敞开型的下水道，孕育着霍乱和痢疾等病菌。直到20世纪中叶，家庭垃圾依旧经常地在露天垃圾场焚烧和处置，垃圾场变成了附近居民的眼中钉，且散发恶臭，老鼠和害虫害兽肆虐其中。化学垃圾则经常随意地留在工业建筑群和处理池的现场。尤其是有毒废弃物可能会被直接填埋，却几乎未采取任何治理措施来阻止废弃物中的有毒物质渗入附近的地表水或污染地下水。

固体废物管理的基本知识

尽管 solid waste, refuse, garbage, trash refuse 的术语经常可以互换，但是固体废物管理方面的专业人员是能够区分它们的。solid waste 和 solid waste, refuse, garbage, trash refuse 是同义词，指任何一种不要的或报废的材料。

被称为固体废物或废物的各种物质可分为如下几类：

严格意义上，Garbage 是指动物或植物废物，具体是指准备膳食时所产生的垃圾。若这些垃圾暴露在自然环境中，会迅速腐烂，散发出恶臭。

Trash 是指不会腐烂降解的固体废弃物（例如：包装袋、瓶子、罐子、建筑材料等）。

有害废物是指具有可燃性、腐蚀性或电抗性（爆炸性的）的废弃物，以及那些含有一定浓度的某些由美国环境保护局认定的有毒化学制品的废弃物。此外，美国环境保护局还有一份列有大约500种其他有害废物的类型清单。虽然联邦法律对大多数工业和商业设施中有害废物的产生、搬运和去除有严格的规定，却不追究家庭生活垃圾中的有害物质的法律责任。

由于大多数人在日常语言中并不区分 garbage 和 trash 这两个词的运用，在此背景下这两个词可以替换。

固体废物流的内容

大多数人都不会花时间去了解他们刚刚扔掉的材料的类型或者垃圾车里面有哪

些东西，如果你问别人垃圾车中占据空间最大的是哪一类材料，你可能会得到许多不同的答案。人们对于固体废物流的构成或特征的理解受到很多因素影响，这些因素包括个人消费、媒体报道以及对乱丢的垃圾和满溢的垃圾桶的直观印象。美国环保局和其他的政府机构定期地汇编我们国家的城市固体废物流数据。下图总结了1996 年美国环境保护局的报告里的重要信息，该报告提供的数据涵盖了根据产品和材料分类的美国城市固体废弃物垃圾的特性。

依据美国环保局报告，城市固体废物流的来源包括住宅、商业、机构和工业场所。（此处的工业废料只包括包装以及行政管理过程中的废弃物，并不包括有害的或者制造过程产生的废弃物。美国环保局报告中也没提到农业废料和城市污泥等其他种类的固体废弃物）

Key to the exercises

I. Translate the following words or phrases into English.
1. solid waste   2. type   3. amount   4. trash; garbage   5. product
6. decompose   7. seep   8. sewer   9. toxic; noxious   10. litter

II. Match the words and expressions with their meanings.
1～5: h i g a b   6～10: c d j f e

III. Match the words and expressions with their meanings.
1～4: c a b h   5～8: g e d f

IV. Translate the following sentences into Chinese.
1. 人类总是产生垃圾并总是以某种方式去处理这些东西。
2. 过去，垃圾总是以最便利的方式被丢弃，很少考虑其对人类健康和环境的影响。
3. 固体废弃物和废物是同义词，指任何一种不要的或报废的材料。
4. Trash 是指不会腐烂降解的固体废弃物（例如：包装袋、瓶子、罐子、建筑材料等）。
5. 依据美国环保局报告，城市固体废物流的来源包括住宅、商业、机构和工业场所。

V. Reading comprehension.
1～5: F T F F T

有害物质

有害物质简介

有害物质是指那些对人及其生存环境构成某种形式的危害的物质。根据《资源保护和回收法案》，有害物质是指会导致或者明显引起死亡率的增长、严重的不可逆的疾病的增加或者不可治愈的可逆性疾病的增加的物质，或者是指当其在不正确的处理、存储、运输以及其他管理过程中会造成对人类健康及其生存环境的显见或潜在的危害的物质。

有害物质的分类

许多有害物质也许不仅归于某一个分类范畴。这些物质根据其形成的危害可分为7种类型：

（1）易燃物——易于点燃且会迅速燃烧的物质。

（2）易爆物/活性物质——当遇到突然的震动、高温或者火源时能够引起突发的、几乎即刻的压力、气体或者热量的释放的爆炸化学品；在一定的震动、压力或者环境温度条件下能导致巨大的化学变化的活性化学品。

（3）致敏物质——人类或者实验性动物首次接触反应很小或者完全没有反应，但反复接触会引起不一定局限于接触部位的明显的反应的物质。皮肤过敏是最常见的反应形式；有几种过敏物质也会引起呼吸道过敏。

（4）腐蚀品——腐蚀品可在接触部位通过化学反应对活组织造成看得见的损伤或者不可逆转的改变。

（5）刺激物——根据浓度和接触时间的长短可在接触部位通过化学反应使活组织产生可逆的炎症的非腐蚀性的化学品。

（6）致癌物质——可导致人类患癌或者因可导致动物患癌而被认为有可能也导致人类患癌的物质。

（7）有毒物质——是指当被摄食、吸入或者通过皮肤吸收时会毒害生物体的物质。

Part III 固体废物管理

Unit 2 固体废物综合治理方法概述

Text 废物综合治理

废物综合治理

废物综合治理系统应结合下面两个以上的流程进行操控：
- 源头减量
- 再利用
- 资源化
- 堆肥
- 焚烧
- 排放
- 填埋

源头减量

在源头就减少进入废物流的垃圾量也许是任何有效的废物综合治理系统中最重要的组成部分。源头减量包括在回收利用或处理之前的可以使固体废弃物体积减小或毒性降低的一切行为活动。

再利用

我们很容易形成"新的东西比旧的好"这种思维定式。遗憾的是，为给新产品让道，每天有许多完好的旧产品被归入到废物流。其中甚至还有些多次使用的也许仍有使用价值的物品——只要把它们维修或翻新一下就行了。

资源化

在美国，每人每天平均产生 4～5 磅的城市生活垃圾。这种垃圾大部分都可以进行资源回收，即让这些材料通过回收和堆肥重新成为可利用的原材料或通过焚化来提供能源。其中最容易回收的材料有纸、金属、玻璃和塑料。人们可以通过手工分选和应用风力分级器、磁铁、旋风分离器、矿石筛、破碎机、研磨机和打包机等各种类型的机械装置重新获得可回收利用的材料。

堆肥

堆肥是将有机废物转化成一种土壤改良剂的被控制的生化过程。堆肥产生一种营养丰富的土壤改良剂，此改良剂也被称为混合肥料，它可通过增加土壤的养分有效性、蓄水能力、曝气程度和生物活性来改善土质（所以植物才会生长）。堆肥除了可实现废物减量和改善土壤外，最近还被发现在防治污染上也有重要用途。在小溪、湖泊、河流堤防或在路边和山坡上实施堆肥可以减少土壤的淤积和流失，还能减少雨水径流中的重金属和有机污染物，防止水污染。堆肥可降解或完全消除像碳氢化合物和杀虫剂这样的污染物。除此之外，使用成熟的堆肥肥料还可以抑制植物病害。

焚化

除了回收利用和堆肥外，来自于城市生活垃圾流中的一些资源可以通过焚烧重新获得。焚化的目的之一是延长填埋场的使用寿命、把气味和卫生问题最小化。使用一个有效的大规模燃烧的焚化炉，可以使固体废弃物体积减量80%～90%或固体废弃物的质量减量65%～75%。

如图1，垃圾焚烧的过程包括运输、存放、焚烧、能源回收、控制排放和残留物处理。

（1）运输——将城市生活垃圾收集和运送到大规模焚烧设施所在地。

（2）存储——将垃圾存放于一个贮藏窖或卸料间。

（3）焚烧——用传送带或吊车将垃圾送进一个料斗，再投进焚化炉焚烧，次级燃烧室则有助于垃圾更完全地燃烧。

（4）能量回收——从燃烧中产生的热量可以转到水管中的水中，让水变成水蒸气。水蒸气则可直接应用到一些工业流程中或被用来发电。

（5）排放控制——应用如静电除尘器和织物过滤器之类的干式和湿式洗涤器和其他空气污染控制设备，可消除排放出来的废气中的一些酸性气体和微粒。

（6）残留物处理——将焚烧后的灰烬和洗涤器及其他污染控制设备上的残渣在填埋场进行填埋处理。

排放

大多数的能量回收设备都采用复杂的焚化控制系统，这些系统被设计用来优化燃烧、尽可能减少待处理的灰烬以及通过减少形成未完全燃烧的产物来提高燃烧效率。这些焚化后残留的废弃物可以以燃烧废气、颗粒物排放、飞灰、底灰等形式排出废物流系统。

填埋

不论多么成功地运行这些流程，其他的废物管理过程都不能完全消除对垃圾填埋场的需要。

现代垃圾填埋场有两类：卫生填埋场和安全填埋场。

人们在卫生填埋场使用过滤器和防渗层来防止土壤污染；使用渗滤液收集和监测系统来防止地下水污染；并使用收集沼气的方法来防止修建填埋场时铺设的隔离膜发生爆裂（图2）。

卫生填埋是可以处于相对良性的环境。一些垃圾填埋场收集垃圾分解产生的甲烷，为邻近社区供给燃料。

必须要经过权威部门许可才能在安全填埋场填埋有毒废弃物，且比卫生填埋场的安全防范措施更严格。不同类型的有毒废弃物被埋在独立的腔室，各个腔室里存放了哪类废弃物都有清楚仔细的存单备案。安全填埋场并非只有一个压实土基底层和可渗水的隔泥网膜（像卫生填埋），而是有两个压实土基底层（黏土层和褐黄色土层）和一层不透水的塑料防渗层来保护垃圾填埋坑的坑壁。顶层覆盖也更深更紧实，还安装有一个地下水监测井帮助检测渗漏情况（见图3）。

安全填埋场在设计和维护方面要求最高。它们是专为无限期存放有毒化学物品和最大警惕度地防止泄漏而设计的。由于埋在那里的化学品的特性和浓度，安全填埋场渗漏将给周边环境带来致命的污染，这比卫生或不卫生填埋场的污染泄漏危害程度要大得多。

Key to the exercises

I. Decide whether each of the following statements is true (T) or false (F) according to the material.

1～5: F T F F T

II. Translate the following words or phrases into English.

1. integrated management   2. source reduction   3. reuse   4. Composting
5. mission control   6. cyclone   7. incinerator   8. fly ash
9. fabric filter   10. secure landfill

III. Match the words and expressions with their meanings.

1～5: e j a c b   6～10: d f g h f

IV. Match the words and expressions with their meanings.

1～5: c e a b d

V. Translate the following sentences into Chinese.

1. 源头减量包括在回收利用或处理之前的可以使固体废弃物体积减小或毒性降低的一切行为活动。
2. 其中最容易回收的材料有纸、金属、玻璃和塑料。
3. 如图1，垃圾焚烧的过程包括运输、存放、焚烧、能源回收、控制排放和残留物处理。
4. 一些垃圾填埋场收集垃圾分解产生的甲烷，为邻近社区供给燃料。
5. 安全填埋场的顶层覆盖也更厚更紧实，还安装有一个地下水监测井帮助检测渗漏情况。

Part IV 科技论文摘要写作

Key to the exercises

I. Put the sentences into right order and find out the key words of the abstract you have formed.

(A): ③①②

environmental management for colleges and universities; ISO 14000EMS; campus waste

(B): ③②①

Ecological environment management; super ministry system; reform

II. Fill in the blanks as follows.

(A): environment criteria; environment standards; transformation mechanism; affecting factors.

to develop a transformation mechanism from EC to ESs suitable for China.

(B): Coatings industry; waste gas; chemometric resolution method; gas chromatography-mass spectrometry

to decrease the requirement of chromatographic resolution and provide the new way of analyzing complex unknown system quickly and precisely.

III. Translate the titles from English into Chinese.

1. 高校环境管理和废物处理处置
2. 西方国家生态环境管理大部制改革及对我国的启示
3. 环境基准向环境标准转化的机制
4. 化学计量学法解析涂料行业废气化学成分

IV. Translate the abstract into Chinese.

环境影响评价在 HSE 管理体系中的应用

摘要：HSE（健康、安全、环境）管理体系已发展成为国际石油石化行业通用的管理体系，中国石油逐步推行 HSE 管理体系，不断探索和实践。文章从环境影响评价与石油企业 HSE 管理体系的相互关联入手，分析将环境影响评价应用于 HSE 管理体系建设的可行性，认为企业的环境管理与环境评价影响评价工作充分结合，以提高企业管理的效率和效益，推动企业长远发展。

V. Translation.

Application and Development of Biological Fluidized-Bed Technology in Wastewater Treatment

Abstract: A history of application and development of biological fluidization technology in waste water treatment is briefly described, while an introduction is made to the operating principles and design features of some new-type biological fluidized-bed reactors developed in recent years. It is pointed out that biological fluidized-bed technology should be developed in such a direction that it will result in lower energy consumption and meet the requirements for the treatment of different waters.

Key words: wastewater; biological treatment; fluidized bed